IntechOpen

Supporting open minds since 2005

Energy Storage Battery Systems - Fundamentals and Applications
http://dx.doi.org/10.5772/intechopen.91100
Edited by Sajjad Haider, Adnan Haider, Mehdi Khodaei and Liang Chen

Contributors
Chris Berry, Himani Gupta, Rajendra K. Singh, Thorsten Hickmann, Toni Adamek, Oliver Zielinski, Thorsten Derieth, Wiqar Hussain Shah, Waqas Muhammad Khan, Asim Ali Yaqoob, Mohamad Nasir Mohamad Ibrahim, Khalid Umar, Venko Beschkov, Elena Razkazova-Velkova, Kudiyarasan Swamynathan, N. Sthalasayanam, M. Sridevi

First published in London, United Kingdom, 2021 by IntechOpen
IntechOpen is the global imprint of INTECHOPEN LIMITED, registered in England and Wales, registration number: 11086078, 5 Princes Gate Court, London, SW7 2QJ, United Kingdom
Printed in Croatia

British Library Cataloguing-in-Publication Data
A catalogue record for this book is available from the British Library

Additional hard and PDF copies can be obtained from orders@intechopen.com

Energy Storage Battery Systems - Fundamentals and Applications
Edited by Sajjad Haider, Adnan Haider, Mehdi Khodaei and Liang Chen
p. cm.
Print ISBN 978-1-83962-906-8
Online ISBN 978-1-83962-907-5
eBook (PDF) ISBN 978-1-83962-915-0

Meet the editors

Dr. Haider is an associate professor in the Chemical Engineering Department, King Saud University, Riyadh, Saudi Arabia. He received his MSc in 1999 and MPhil in 2004, both from the Institute of Chemical Sciences, University of Peshawar, KPK, Pakistan. He obtained a Ph.D. in 2009 from the Department of Polymer Science and Engineering, Kyungpook National University, Taegu, South Korea. His research work focuses on the development of biopolymer composites, polymer hydrogel, and the preparation of electrospun nanofibers in biomedical applications and for the removal of hazardous materials from an aqueous medium.

Dr. Adnan Haider is an assistant professor at the Department of Biological Sciences, National University of Medical Sciences (NUMS), Pakistan. He received his MSc from Kohat University of Science and Technology, Pakistan, and obtained an MS leading into a Ph.D. from Kyungpook National University, South Korea. Dr. Haider completed a post-doctorate at Yeungnam University, South Korea. His research work focuses on the development of scaffolds for tissue regeneration, biopolymer composites, polymer hydrogel, drug delivery systems, and preparation of electrospun nanofibers, and assessment of their potential use in biomedical applications and removal of hazardous materials from aqueous medium.

Dr. Mehdi Khodaei received his bachelor's degree in 2005 and a master's degree in 2008 from the Department of Materials Engineering, Isfahan University of Technology, Iran. In 2014, he obtained his Ph.D. in Nanotechnology-Materials Science from the University of Tehran, Iran. He was a visiting researcher at Pohang University of Science and Technology (POSTECH), Korea from March 2012 to June 2013. His research area is functional nanomaterials. He is currently working as an assistant professor at the Faculty of Materials Science and Engineering, K.N. Toosi University of Technology, Iran.

Dr. Liang Chen is an associate professor at Ningbo Institute of Materials Technology and Engineering, Chinese Academy of Sciences (CAS). His research interests focus on electrochemical energy storage and conversion. He received his Ph.D. in Physical Chemistry from Lanzhou Institute of Chemical Physics of CAS in 2008 and completed a three-year postdoctoral training at the National Research Council Canada (2008–2011). Dr. Chen is the author and co-author of more than thirty peer-reviewed journal articles. He served as a member of the Youth Innovation Promotion Association (CAS) in 2017 and was awarded the Zhuliyuehua Outstanding Doctor Award in 2008.

Contents

Preface

This book discusses the scientific and technical principles underpinning the major energy storage technologies, including traditional and new battery systems, such as lithium, redox flow, and regenerative batteries as well as bio-electrochemical processes. It discusses the significant advancements that have been achieved in the development of methods and materials for various storage systems.

The first section discusses the higher power regenerative battery discharger for nuclear power plants and provides information on energy storage and conversion devices, integration of devices and systems to communicate, recent progress on current technologies, and ideas for the next generation. This section also discusses redox flow batteries with emphasis on graphite-filled thermoplastic-based composites. This composite is a suitable material for bipolar plates in redox flow battery applications. Unlike metals, composite plates can provide exceptional resistance to the highly aggressive chemical environment at elevated temperatures in combination with an electrochemical potential in battery operation.

The second section discusses the present structure of the lithium-ion supply chain, with an emphasis on raw material production, as well as the need for and problems connected with getting enough financing in an industry that has not seen rapid expansion. This section also details the role of electrolytes in Li-ion batteries. Most of the electrolytes utilized minimize internal resistance. However, they are not electrochemically stable and have caused serious issues such as electrode reactivity, dissolution of electrode ions, leakage, volatility, and rapid Li dendrite development. Hence an ionic liquid (IL)-based polymer electrolyte is used over liquid electrolyte in which the IL acts as a plasticizer and improves ionic conductivity and amorphicity. Because of their excellent thermal and electrochemical stability, these electrolytes can be utilized in high-voltage Li batteries. Furthermore, their mechanical stability aids in the suppression of Li dendrite development. As a result, polymer electrolytes may pave the way for a new era in battery applications.

The third and final section addresses the concepts, accomplishments, and potential trends in energy generation by microbial fuel cells (MFCs), as well as processes in biotechnology and wastewater treatment by microbial electrolysis cells. MFCs have received a lot of attention due to their moderate operating conditions and use of a range of biodegradable substrates as fuel to create a lot of energy or efficiently treat pollutants. To demonstrate their performance, the issues faced by the electrode materials need to be addressed. Hence this section also discusses electrode materials and their functions.

This book is useful for students and teachers alike for its thoughtful account of core concepts.

Sajjad Haider
Department of Chemical Engineering,
King Saud University,
Riyadh, Saud Arabia

Adnan Haider
Department of Biological Sciences,
National University of Medical Sciences,
Rawalpindi, Punjab, Pakistan

Mehdi Khodaei
Faculty of Materials Science and Engineering,
K.N. Toosi University of Technology,
Tehran, IRAN

Liang Chen
Advanced Lithium battery Engineering Lab,
Ningbo Institute of Materials Technology and Engineering,
Chinese Academy of Sciences,
Ningbo, China

Section 1

Power Generation Batteries

Chapter 1

Design of High Power Regenerative Battery Discharger System for Nuclear Power Plant

Kudiyarasan Swamynathan, N. Sthalasayanam and M. Sridevi

Abstract

In a Nuclear Power reactor, safety loads are backed by standby battery system. The healthiness of the battery is very essential requirement and prominent attention is given to availability and reliability of battery supply in nuclear plants. Hence regular monitoring and testing the performance of the battery is a prime requirement. The capacity and load cycle discharge testing of the battery is done annually and the current system employed is to discharge the battery current through resistor banks, which results in unusable power consumption and is uneconomical. The growing trend in power electronics field has given the new technology of regenerating the dissipated power to grid. This paper proposes a high power electronic regenerative technology with high efficiency, low harmonics to pump the dc power to the grid. Though, it is available at lower rating in industry, the paper proposes a high power regenerative discharge system. The topology selected is interleaved boost converter interfaced to a three phase grid connected inverter. The challenges involved are high power operation, steep current discharges with a minimal interference to the normal plant operation power supplies during the regeneration. This paper also presents the system design and simulation results.

Keywords: Regenerative battery discharger, Interleaved Boost Converter, Grid connected Inverter, harmonics reduction, nuclear safety load

1. Introduction

In nuclear power plants, the batteries and DC power system plays a prominent role in the reactor safety and shutdown. Hence the preventive maintenance and testing the capability of battery to cater the safety loads during power failure are done regularly as per nuclear standards. There is requirement of performing C10 and HRD discharging the batteries. The growing trend of large capacity NPP's being installed worldwide has increased the battery ratings substantially, resulting the discharge testing requirements to high current and voltage levels [1]. In conventional plants, resistor banks employed for discharging battery has resulted in power loss, hence there is a need to use the regenerate power electronics system in the battery application. Even though Low power discharge systems are available in the current scenario, the discharging of high power battery systems for higher currents and shorter duty cycles is the motivation behind this paper [2, 3]. The system employs two phase interleaved boost converter to step up the connected battery

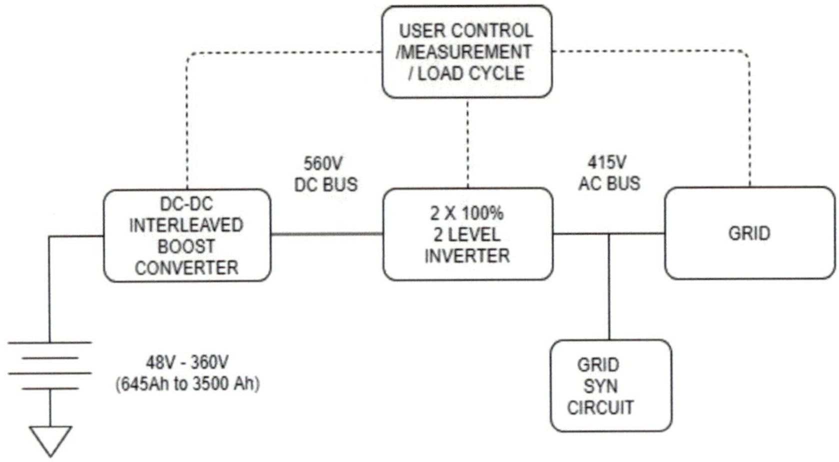

Figure 1.
Block diagram of grid connected regenerative discharger system.

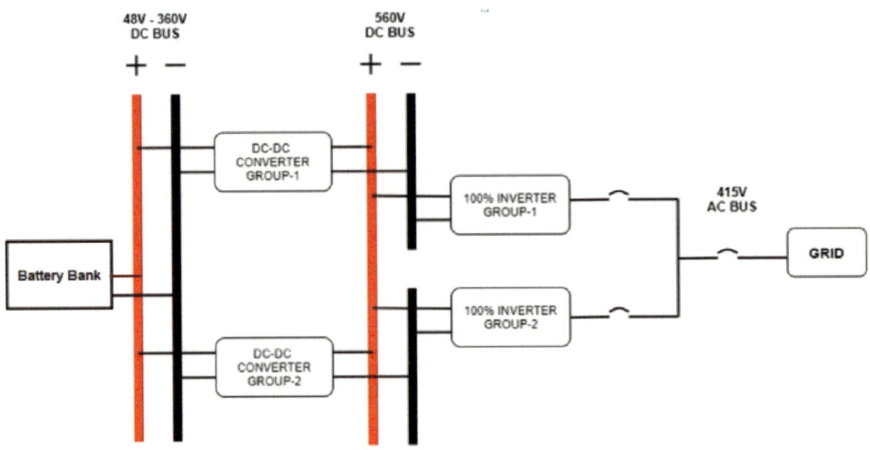

Figure 2.
Schematic of regenerative battery discharge system.

voltage level from (48–360)V DC to 560 V DC as shown in **Figure 1**. The boost converter topology is selected to ensure minimum DC ripples in the inverter dc bus.

The boost converter is modularized into two independent 50% power units and they are independently connected to 2 X 50% two level 415 V Inverter.

Accordingly the boost converter is designed in the modular approach, with two sets of 3 X 100A boost converter to provide required current rating as shown in **Figure 2**. The Grid connected inverter discharges the power to Grid by converting DC to AC. The output isolation transformers in the inverter reduce the third harmonic component and also provide isolation between grid and the plant system.

2. Regenerative power electronics topology

2.1 High power load cycles of batteries

The plant consist of the following battery banks for the reactor safety loads in the range of 48 V DC, 220 V DC, 360 V DC. Following are the various battery bank capacities with various DC voltage ranges as shown in **Table 1**.

In the nuclear power plant the load cycle is more complex when compared with other conventional plants [4]. A typical example of load cycle is given in **Table 2**.

2.2 DC-DC converter topology

The topology selected for our application is dual phase Interleaved Boost Converter (IBC). Interleaving is a method of multi-phasing in which two converters are connected in parallel. In interleaved boost converters, the number of phases has a significant impact on the current ripple [5, 6]. Though ripple content reduces with increase in the number of phases, the power circuit, on the other hand, the complexity of the circuit and triggering signals will be increase [7].

In this paper, a two-phase interleaved boost converter is selected as DC-DC Converter topology [8]. In a two-phase converter, there are two Output stages that are driven 180 degrees out of phase as shown in **Figure 3**. By splitting the current into two parallel paths, conduction losses can be reduced, leading to improved efficiency compared to a single-phase converter. The ripple generated from switch S1 and complimentary switch S2 cancels each other [9, 10]. Employing coupled inductors in this topology adds to the advantage of the input current-ripple cancelation from magnetic coupling between the phases. The frequency of the current ripple is twice for two phase IBC than the conventional boost converter [11]. The

Voltage (V)	Ampere hour (Ah)	No. of sets
48 V	800 Ah	4
48 V	1600 Ah	4
220 V	650 Ah	4
220 V	2400 Ah	2
220 V	3500 Ah	2
360 V	1900 Ah	4

Table 1.
Various batteries with different voltage rating and capacity available in the plant.

Discharge current duration (minutes)	Battery bank voltage (volts) and capacity (Ah)					
	48 V 800 Ah	**48 V 1600 Ah**	**220 V 650 Ah**	**220 V 2400 Ah**	**220 V 3500 Ah**	**360 V 1900 Ah**
	Current (Amps)					
0–1	200	500	350	600	600A for 240 min	522A for 60 min
1–10	200	500	100	290		
10–15	200	500	150	290		
15–16	200	500	150	590		
16–30	200	500	150	590		
30–60	50	100	50	590		
60–240	50	100	50	20		
240–840	25	50	10	20		
840–841	—	—	20	50		

Table 2.
The load demanded with required time duration of all battery banks available in the plant.

converter must be able to operate over a wide input-voltage range (40-400 V) to accommodate batteries of different voltages (48 V, 220 V, 360 V). Because of the wide input range, the converter also must be able to operate with a wide input-voltage to output-voltage ratio [12].

The controller sets the pre-set duty cycle as input for converter switching as the input voltage to converter is selected by the operator. The output voltage of the boost converter is fixed to 560 VDC for providing required DC bus voltage for the inverter module. The main design consideration of this converter is done with respect to the battery's end bank voltage [13]. During discharging of batteries, as the battery reaches the end bank voltage, the voltage boosting has to be done by the converter for the reduced input voltage to maintain a steady DC bus at inverter input and has to supply the rated current at the output. This design consideration is implemented through dynamic duty cycle variation based on the input DC voltage feedback to the converter [14, 15]. Switching frequency of the converter is selected nominally at 10 kHz and the duty cycle for the switching is selected as per the input-voltage equations of a traditional boost converter. The inductors selected for the converter is uncoupled type.

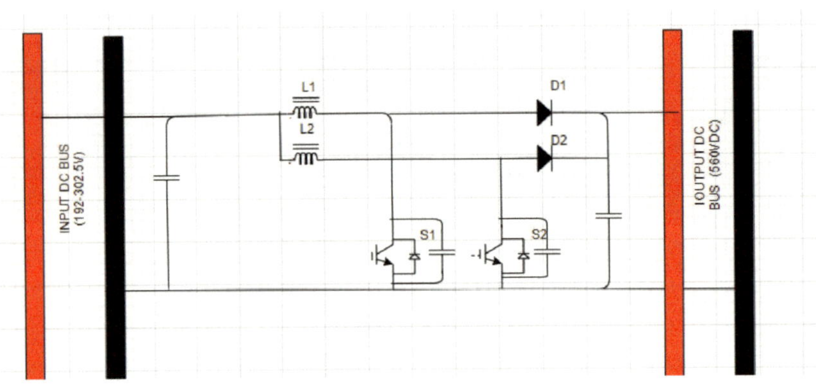

Figure 3.
Schematic of boost converter.

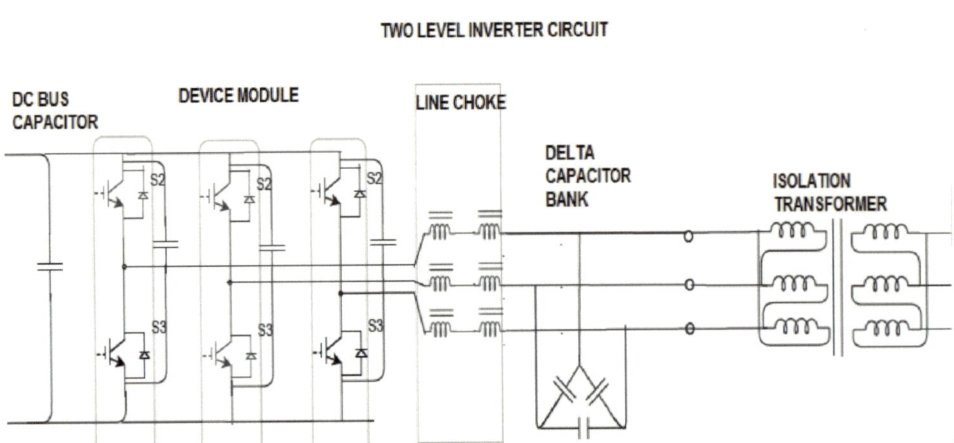

Figure 4.
Schematic of two level inverter systems.

2.3 Grid connected inverter topology

2 X 100kVA IGBT (Insulated Gate Bipolar Transistor) based inverter at the DC boost converter output operates in synchronism with Grid supply.

Two power stacks for each phase is designed for effective load sharing and thereby reduced heat dissipation [16]. The switching frequency of inverter is selected at 1 kHz. Sinusoidal PWM algorithm is implemented for generating inverter switching pulses. There are two inverters each rated 50% capacity (100KVA) connected parallel sharing the load. Inverter-1 is fed from Group-1 DC Boost converter and Inverter-2 is fed from Group-2 Boost converter. In case of failure of one inverter, the 50% load can be taken up by the second inverter. The topology of two level inverter is selected to reduce the complexity in inverter design. The output harmonics primarily 5th and 7th harmonic components are reduced with the help of LC Filter as shown in **Figure 4**. The 415/415 V output isolation Δ-Δ transformer eliminates third harmonics in the output. The inverter is in synchronized to grid during operating conditions. The battery power is delivered to grid with the help of inverter synchronized to grid supply.

3. Circuit parameter design

The design of interleaved boost converter is very similar to traditional boost converter design.

3.1 Duty cycle (D)

Generally output voltage equation of any conventional boost converter is given in (1), Duty cycle for any input and output voltages can be represented as (2),

$$\mathrm{V_O} = \frac{Vd}{(1-D)} \tag{1}$$

$$\mathrm{D} = \frac{(\mathrm{Vo–Vd})}{\mathrm{Vo}} \tag{2}$$

Boost converter to work with three different input voltages 48 V, 220 V, and 360 VDC respectively as represented below,

$D_{48V} = D_{max}$ = (560–48) / 560 = 0.914.
D_{220V} = (560–220) / 560 = 0.607.
$D_{360V} = D_{min}$ = (560–360) / 560 = 0.357

3.2 Current ripple (ΔI_O)

Each Boost Converter is designed for 30KW Power rating.
Load current I_O = Output power /Output voltage
$I_O = 33 \times 10^3 /560 = 59A$.
For D < 0.5

$$Iorms = \frac{\mathrm{Io}\sqrt{\mathrm{D}(1-\mathrm{D})}}{2(1-D)} \tag{3}$$

For D > 0.5

$$Iorms = \frac{\mathrm{Io}\sqrt{\frac{1}{2}(2D-1)(2-2\mathrm{D})}}{2(1-D)} \tag{4}$$

By using (3) and (4), Output rms current is arrived for all modes as below.
Iorms (48 V) = 91.52 A.
Iorms (220 V) = 21.72 A.
Iorms (360 V) = 21.98 A.

Considering ΔI_O load current ripple to be 8% of output current. For evaluation, maximum duty cycle, in this case for input 48 VDC Input mode, current ripple (ΔI_O) is arrived to be 7.2 A. For other duty cycles also, ΔI_O can be arrived in similar lines [17].

3.3 Inductor value

For device switching frequency set at 10 kHz. The inductance parameter can be calculated as below.
Switching Time Ts = 1/fs.
= 1/10 kHz = 100 µs.

$$\mathrm{L} \geq \frac{(V_{in}D_{max}T_s)}{2\Delta I_o} \tag{5}$$

For Vin = 48 V & D max = 0.914.
Substituting the values of Vin & D max in (5),
$\mathrm{L} \geq (48)(0.914)(10^{-4})$ / 14.
L ≥ 313 µH.

The inductance parameter is selected to be greater than 313 µH, so the optimized design value is 368 µH considering design tolerances.

3.4 Capacitance value

The capacitor selection is decided based on voltage ripple at output. Considering ΔV_O output voltage ripple to be 1 V.
For D = 0.914.

$$\mathrm{C} \geq \frac{(V_O\mathrm{D})}{\mathrm{R}\,\Delta V_O F_s} \tag{6}$$

Substituting the values of V_o *&* D_{max} in (6),

C = (560) (0.914) / $(100^*1^*10^5)$.
C ≥ 102 µF.
For D = 0.357.
C ≥ 199 µF.
For D = 0.607.
C ≥ 45 µF.

The capacitance parameter is selected to be greater than 200 µF. considering 5% margin, capacitance value arrived at 210 µF.

4. Simulation studies

The Boost converter circuit and two level Inverter circuit were simulated with the PLECS & PSIM power electronics simulation tools. The simulation parameters were decided based on the design parameters and various input voltage selections as tabulated in **Table 3** [18].

The Schematic of converter circuit for all the input voltages are designed and typical 48 V DC Input simulation circuit is shown in **Figure 5** [19]. The switching patterns of two IGBTs (S1, S2) used in the two phase IBC is shown in **Figure 5** [20]. The phase shift between the two phase limbs is 180 degrees.

For all the three modes and the output of 560 V DC was achieved and corresponding output voltage, inductor currents and input current were obtained graphically [21]. The 48 V Input DC circuit is simulated for duty cycle of 0.9142 and the results are represented in **Figure 6** [22].

The output voltage obtained for 48 V DC input circuit in simulation has very low ripple voltage content within 1 Volt [23]. The inductor current ripple is 30% and the input current ripple is 15%, which are well within design limits. Similarly, the duty cycle 'D' is varied for input voltages 220 VDC and 360 V DC as per the **Table 3** and the corresponding graphical results are shown in **Figures 7** and **8** [24].

The simulation results show very less ripple in input current and voltage compared to any conventional boost converter [25]. The Inverter design parameters were arrived with conventional methods, output LC Filter is designed for reducing higher order harmonics at output current and smoothening the output current. The simulation parameters are tabulated in **Table 4**.

Parameter	Input voltages		
	48 V	220 V	360 V
L1/L2	368 μH	368 μH	368 μH
C	210 μF	210 μF	210 μF
Duty Cycle	0.9142	0.607	0.357
Switching Frequency	10KHz	10KHz	10KHz

Table 3.
Simulation parameters for interleaved boost converter.

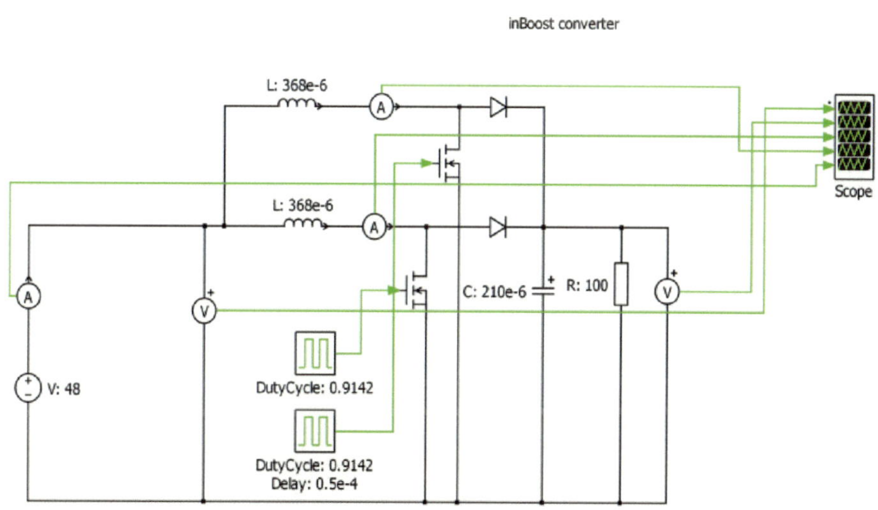

Figure 5.
Schematic of interleaved boost converter with 48 V input.

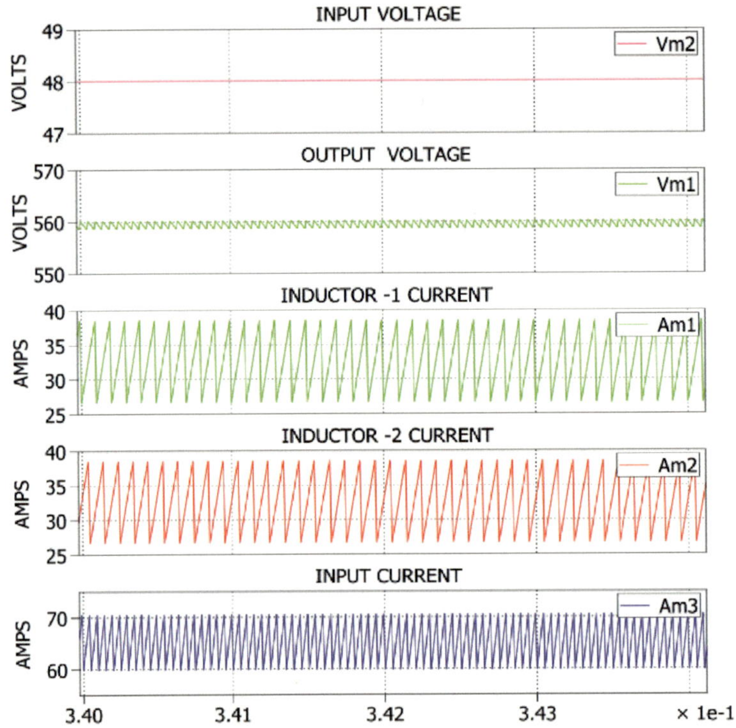

Figure 6.
Waveforms of interleaved boost converter with 48 V input.

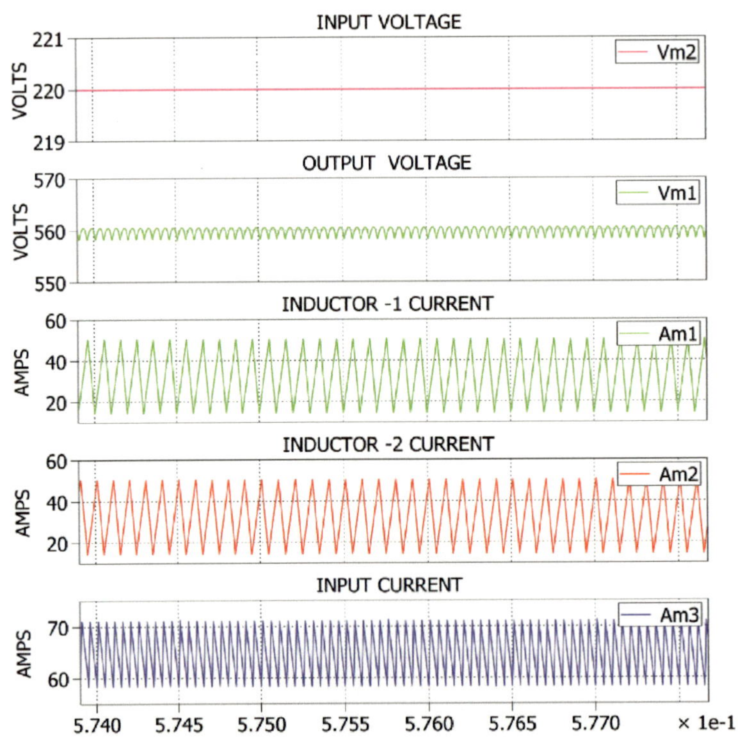

Figure 7.
Waveforms of interleaved boost converter with 220 V input.

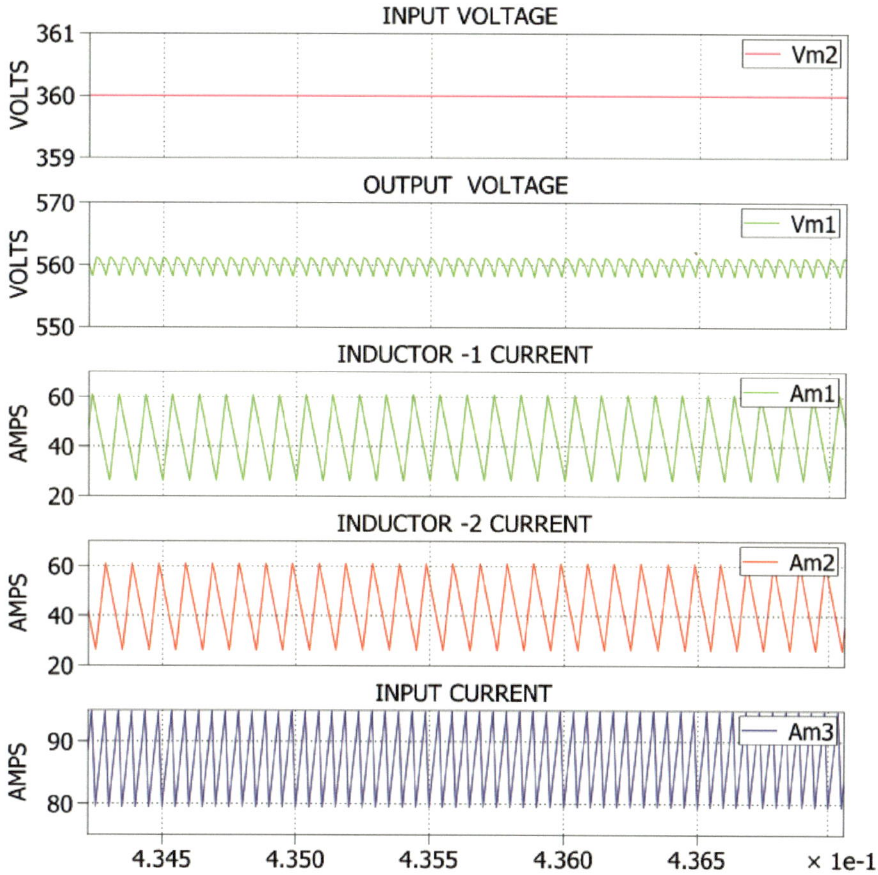

Figure 8.
Waveforms of interleaved boost converter with 360 V input.

Parameter	Values
Dc Bus Input	560 VDC
Output 'C'	132 μF
Output 'L' Choke	380 μH
Δ-Δ Transformer	415/415 V
Switching Frequency	1KHz
Modulation Index 'M'	0.8

Table 4.
Simulation parameters for inverter.

The Schematic of inverter simulation circuit with the PWM Generators is given in **Figure 9** and corresponding output voltage waveforms as obtained from simulation is shown in **Figure 10** [26, 27].

The FFT of output AC Voltage is shown in **Figure 11**. **Figure 11** also shows harmonic spectra in the waveform. The predominant harmonics in two level inverters are 5th and 7th harmonics, which is reduced by incorporating LC Filter at output [28]. The Δ-Δ Isolation Transformer at output gives circulating path for third harmonics zero sequence currents and thereby resulting very less third

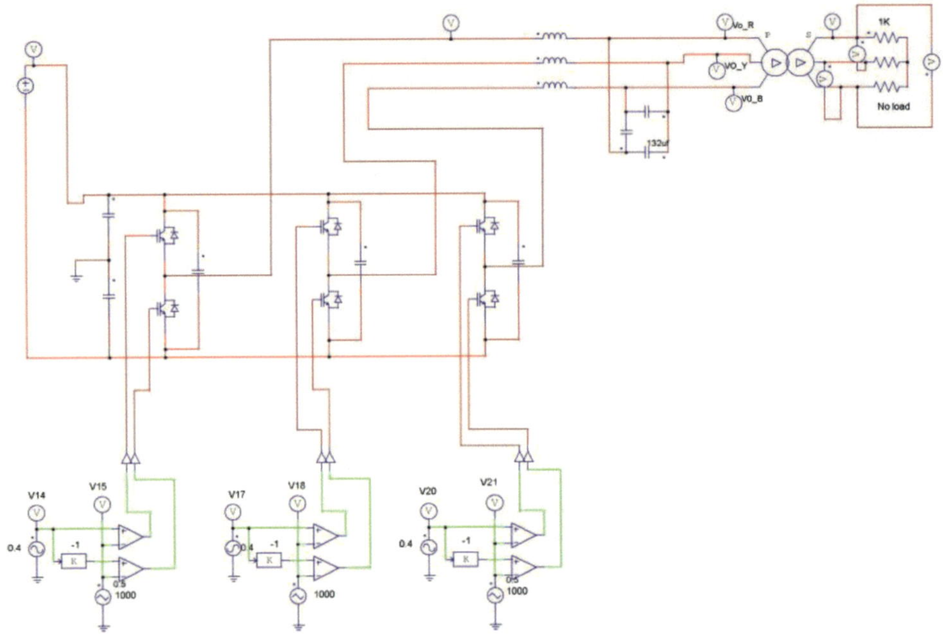

Figure 9.
Simulation of two level inverter systems

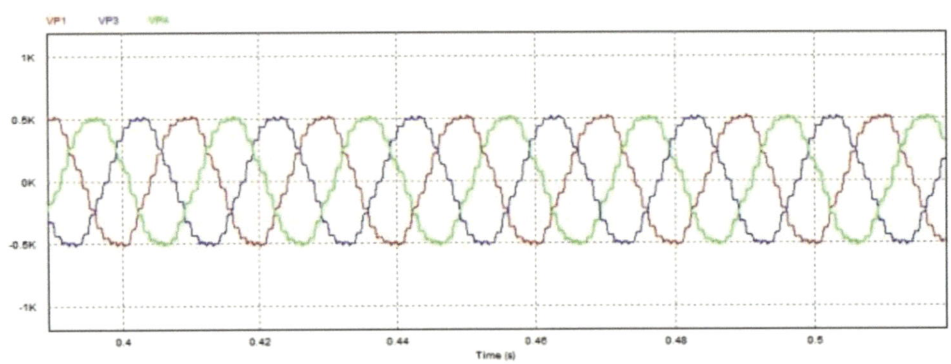

Figure 10.
Waveform of inverter O/P voltage.

harmonic output to load [29]. THD at output is calculated to be 7.2% as tabulated in **Table 5** [30].

5. Design parameter verification

All Design parameters were successfully verified in our simulation and the verification table is given in **Table 6**.

6. Challenges in simulation of high power converter

The very advantage of using interleaved converters for reducing the current and voltage ripple resulted in low design value of capacitances. However incorporating

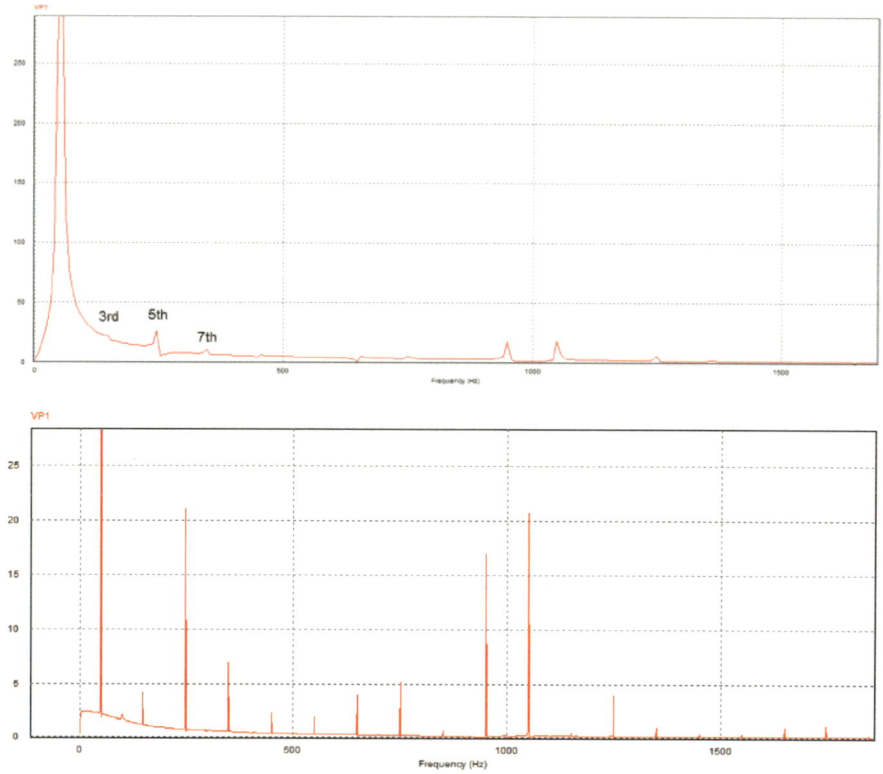

Figure 11.
FFT output of inverter output voltage.

Harmonic number	Percentage %
3rd	1 .2
5th	7.36
7th	5.47
9th	5.35
THD	7.2

Table 5.
Percentage harmonics in output voltage.

Parameter	Input voltages		
	48 V	220 V	360 V
BOOST CONVERTER			
Input Current Ripple	17%	20%	21%
Output voltage	560 V	560 V	560 V
Output voltage ripple	0.8 V	0.8 V	0.8 V
Duty Cycle	0.914	0.607	20.357
INVERTER			
Output voltage	415 V ± 5%		
Output current	150A		
THD	7.2%		

Table 6.
Design parameter verification of regenerative discharger.

two large inductances in both the switch stacks results in a real challenge in design as well as for simulation models. The simulation parametrization were calculated and designed for continuous conduction mode of inductor currents.The uncoupled inductor model is taken up for reducing the complexity in simulation.

7. Conclusion

The periodic testing and discharging the batteries to ensure the capacity and availability of battery systems is mandatory in nuclear power plants. The batteries of large power capacity are discharged with the regenerative technology to the grid by utilizing this technique. The design intents of lower current ripple and output voltage ripples is realized with the interleaved boost converter topology. The Harmonics in AC Output is well minimized with the help of optimum LC Filter design. This paper is conceptual base for designing high power battery regenerate discharge system without disturbance to normal plant operation with grid. A 100A Boost converter prototype model was assembled and was paralleled with 6 similar units for 600A Current capacity. Thus, the proposed design has a huge capacity to deliver energy savings, output voltage and current regulations with very minimum net harmonics for wide range of discharge currents and voltages.

Acknowledgements

The authors sincerely thank Bharatiya Nabhikiya Vidyut Nigam Limited (BHAVINI), Department of Atomic Energy, Kalpakkam, India to carry out this research.

Conflict of interest

No conflict of interest.

Author details

Kudiyarasan Swamynathan[1*], N. Sthalasayanam[2] and M. Sridevi[1]

1 Bharatiya Nabhikiya Vidyut Nigam Limited, Department of Atomic Energy, Kalpakkam, Tamilnadu, India

2 Master of Engineering, SRMIST, Chennai, India

*Address all correspondence to: kudiyarasan@rediffmail.com

References

[1] Mahdiyeh Khodaparastan, Ahmed A. Mohamed, and Werner Brandauer, "Recuperation of Regenerative Braking Energy in Electric Rail Transit Systems" IEEE Transactions on Intelligent Transportation Systems, Vol. 20, No. 8, pp. 2831-2847, Aug. 2019.

[2] N. KrishnaKumari, D.S.G.Krishna, M. PrashanthKumar, "Transformer less high voltage gain step-up DC-DC converter using Cascode technique", Energy Procedia., Vol. 117, pp. 45-53, Jun. 2017. Doi:10.1016/j.egypro.2017.05.105

[3] Ajit T N, "Two Stage Interleaved Boost Converter Design and Simulation in CCM and DCM", International Journal of Engineering Research & Technology., Vol. 3, Issue 7, Jul. 2014.

[4] Lucas S. Xavier, William C. S. Amorim, Allan F. Cupertino1, Victor F. Mendes, Wallace C. do Boaventura and Heverton A. Pereira, "Power converters for battery energy storage systems connected to medium voltage systems: A comprehensive review", BMC Energy, Article No. 7, 2019. [Online]. Available: https://doi.org/10.1186/s42500-019-0006-5

[5] W. Subsingha and P. Sarakarn, "4 Phase Interleaved DC Boost Converter for PEMFC Applications", Procedia Engineering., Vol. 32,pp.1127–1134,2012[Online]. Available: https://www.sciencedirect .com/science/article/pii/S1877705812014907.

[6] Saowanee Kanta, Boonyang Plangklang, WanchaiSubsingha, "Design of a Bi-directional DC-DC 4 phase Interleave converter for PV applications", Energy Procedia,Vol. 56, pp. 604-609, 2014. doi: 10.1016/j.egypro.2014.07.199.

[7] Haiping Xu, Xuhui Wen, Ermin Qiao, Xin Guo, Li Kong, "High Power Interleaved Boost Converter in Fuel Cell Hybrid Electric Vehicle", IEEE International Conference on Electric Machines and Drives., pp. 1814-1819 May. 2005. doi:10.1109/IEMDC.2005.195966.

[8] R. Seyezhai and B. L. Mathur, "A Comparison of Three-Phase Uncoupled and Directly Coupled Interleaved Boost Converter for Fuel Cell Applications", International Journal on Electrical Engineering and Informatics., Vol. 3, No. 3, pp. 394-407, 2011

[9] WeerachatKhadmun and WanchaiSubsingha, "High Voltage Gain Interleaved DC Boost Converter Application for Photovoltaic Generation System", 10th Eco-Energy and Materials Science and Engineering, Energy Procedia., Vol. 34, pp. 390-398, 2013. doi: 10.1016/j.egypro.2013.06.767.

[10] K. Latha Shenoy, C.Gurudas Nayak, Rajashekar P Mandi, "Design and Implementation of Interleaved Boost Converter", International Journal of Engineering and Technology (IJET)., Vol. 9.,No. 3S., pp. 496-502., July 2017. [Online]. Available: DOI: 10.21817/ijet/2017/v9i3/ 170903S076.

[11] Aydin Boyar, ErsanKabalci, "Comparison of a Two-Phase Interleaved Boost Converter and Flyback Converter", IEEE 18th International Power Electronics and Motion Control Conference., pp 352-356, Aug. 2018. doi:10.1109/EPEPEMC.2018.8521891

[12] SairatunNesaSoheli, Golam Sarowar, Md. AshrafulHoque, MdSaidul Hasan, "Design and Analysis of a DC - DC Buck Boost Converter to Achieve High Efficiency and Low Voltage Gain by using Buck Boost Topology into Buck Topology", International Conference on Advancement in Electrical and Electronic Engineering., p-094,pp. 1-4., Nov. 2018.

[13] Yan Cao, Yiqing Li, Geng Zhang, KittisakJermsittiparsert, Maryam Nasseri, "An efficient terminal voltage control for PEMFC based on an improved version of whale optimization algorithm", Energy Reports., Vol. 6, pp. 530-542, 2020. [Online]. Available: https://doi.org/10.1016/j.egyr.2020.02.035.

[14] B. Salhi, T. Ahmed-Ali, H. El Fadil, E. Magarotto, F. Giri,"Adaptive Output Feedback Control of Interleaved Parallel Boost Converters", 11th IFAC International Workshop on Adaptation and Learning in Control and Signal Processing., pp. 311-317, July 2013. [Online]. Available:https://www.sciencedirect.com/science/article/pii/S1474667016329639

[15] H. El Fadil, F. Giri, J.M. Guerrero, M. Haloua, A. Abouloifa, "Advanced Control of Interleaved Boost Converter for Fuel Cell Energy Generation System", Proceedings of the 18th World Congress The International Federation of Automatic Control Milano., pp. 2803-2808, Aug.–Sep.2011,[Online]. Available: https://www.sciencedirect.com/ science/article/pii/S1474667016440383

[16] Marcos B. Ketzer, David C. C. Freitas, Antonio M. N. Lima, Alexandre C. Oliveira,"Battery Test System with Regenerating Capability", EngOpt 2016 - 5th International Conference on Engineering Optimization, Brazil., 19-23 June 2016.

[17] J. DivyaNavamani, K. Vijayakumar, R. Jegatheesan, "Non-isolated high gain DC-DC converter by quadratic boost converter and voltage multiplier cell", Ain Shams Engineering Journal, Vol. 9, pp. 1397–1406, 2018. [Online]. Available: http://dx.doi.org/10.1016/j.asej.2016.09.007.

[18] Ahmad Alzahrani, PouryaShamsi, Mehdi Ferdowsi, and Cihan H. Dagli, "Chaotic Behavior in High-Gain Interleaved Dc-dc Converters", Procedia Computer Science., Vol. 114, pp. 408-416, 2017. doi: 10.1016/j.procs.2017.09.002

[19] Nandakumar Selvaraju, Prabhuraj Shanmugham, Sakda Somkun, "Two-Phase Interleaved Boost Converter Using Coupled Inductor for Fuel Cell Applications", Energy Procedia., Vol. 138, pp. 199-204, Oct. 2017. doi:10.1016/j.egypro.2017.10.150

[20] Slavomir Kascak, Michal Prazenica, Miriam Jarabicova, Marek Paskala, "Interleaved DC/DC Boost Converter with Coupled Inductors", Power Engineering And Electrical Engineering., Vol. 16, No. 2, Jun. 2018. doi: 10.15598/aeee.v16i2.2413.

[21] SiriwatSakulchotruangdet, SudaratKhwan-on, "Three-Phase Interleaved Boost Converter with Fault Tolerant Control Strategy for Renewable Energy System Applications", International Electrical Engineering Congress, Procedia Computer Science., Vol. 86, pp. 353 – 356, 2016. doi: 10.1016/j.procs.2016.05.095.

[22] WalidEmar, ZayedHuneiti , SofyanHayajneh, "Analysis, Synthesis and Simulation of Compact Two-channel Boost Converter for Portable Equipments Operating with a Battery or Solar Cell", International Conference on Communication, Management and Information Technology, Procedia Computer Science., Vol. 65, pp. 241-248, 2015. doi: 10.1016/j.procs.2015.09.118.

[23] S.Kalaimaran,SriB.Revathi,M. Prabhakar, "High Step-Up DC-DC Converter with Reduced Switch Stress and Low Input Current Ripple", 1st International Conference on Power Engineering, Computing and Control., Vol. 117,pp. 1182-1189, 2017. Doi: 10.1016/j.egypro. 2017.05.245.

[24] MiriamJarabicová, "The Parametric Simulation of the Interleaved Boost Converter for the Electric Transport Vehicle", 13th International Scientific Conference on Sustainable, Modern and Safe Transport , Transportation Research Procedia., Vol. 40, pp. 287-294. 2019. doi:10.1016/j.trpro.2019.07.043.

[25] Shri Soundharya J, Sowmiya A, Subhitcha R, Dr. R. Seyezhai, "Performance Evaluation of Interleaved Boost Converter Topologies for Photovoltaic Applications", International Journal of Pure and Applied Mathematics., Vol. 118, No. 24, 2018.

[26] T. J. Liang, T. Wen, K. C. Tseng, and J. E Chen, "Implementation of a Regenerative Pulse Charger Using Hybrid Buck-Boost Converter", 4th IEEE International Conference on Power Electronics and Drive Systems. IEEE PEDS 2001 - Indonesia. Proceedings, pp.437-442, 2001.

[27] NaciGenc, IresIskender, "An improved soft switched PWM interleaved boost AC–DC converter", Energy Conversion and Management., Vol. 52, pp. 403-413, 2011. doi:10.1016/j.enconman.2010.07.016.

[28] Sheng-Yu Tseng, Chih-Yang Hsu,"Interleaved step-up converter with a single-capacitor snubber for PV energy conversion applications", Electrical Power and Energy Systems., Vol. 53, pp. 909–922, 2013. [Online]. Available: http://dx.doi.org/10.1016/j.ijepes.2013.06.007.

[29] Yiwei Ma, Kai Sun, Xiaonan Lu, Lipei Huang, Seiki Igarashi, "A Grid-Connected Hybrid Cascaded H-Bridge Inverter", International Conference on Electrical Machines and Systems., 20-23 Aug. 2011., DOI: 10.1109/ICEMS.2011.6073726

[30] GionataCimini, Maria Letizia Corradini, GianlucaIppoliti, Giuseppe Orlando, Matteo Pirro, "Passivity-Based PFC for Interleaved Boost Converter of PMSM drives", 11th IFAC International Workshop on Adaptation and Learning in Control and Signal Processing., July 2013. Doi:10.3182/20130703-3-FR-4038.00128.

Chapter 2

Key Components in the Redox-Flow Battery: Bipolar Plates and Gaskets – Different Materials and Processing Methods for Their Usage

Thorsten Hickmann, Toni Adamek, Oliver Zielinski and Thorsten Derieth

Abstract

Graphite filled thermoplastic based composites are an adequate material for bipolar plates in redox flow battery applications. Unlike metals, composite plates can provide excellent resistance to the highly aggressive chemical environment at elevated temperatures in combination with an electrochemical potential in battery operation. The chapter therefore gives an overview of the most important requirements for the graphite-plastic composite material and thus also for the bipolar plates, as well as the different characterization methods of the bipolar plates. In the following, both the modern composite materials based on polypropylene (PP) and polyvinylidene fluoride (PVDF) and their general properties are described with a focus on improved long-term stability. Furthermore, recycling is also considered. One section is dedicated to seals, which - as so often - are an underestimated component of redox flow batteries. In this gasket part of the chapter, the most common materials and interactions between gaskets and other stack components are presented, as well as the material properties, characterization and processing methods of the gaskets.

Keywords: bipolar plate (BPP), gasket, graphite compound, composite, graphite plate, polypropylene (PP), polyvinylidene fluoride (PVDF), ethylene-propylene-dien-monomer (EPDM), fluoroelastomer (FKM)

1. Introduction

Redox flow batteries (RFB) are electrochemical reactors suitable for storing electrical energy by chemical reactions [1]. Depending on the technology used, this reaction can take place at elevated temperatures and/or in aggressive media, with an electrochemical potential superimposed. In recent years, the technical requirements on materials and components of the reactor of the Redox flow battery have therefore become more and more demanding. The battery unit consists of many stacked cells which are connected in series to a Flow battery stack. Each cell in turn consists of various components such as the proton exchange membrane, seals, frames and the conductive bipolar plate which provides the connection from cell to cell up to the end of the stack where the generated current is collected.

RFBs, in particular vanadium redox flow batteries (VRFBs), have now reached a considerable degree of technical maturity and the systems are available on the market through many suppliers. However, due to a high remaining cost structure - partly due to a lack of economies of scale - the profitable market introduction of flow batteries still suffers from a high market acceptance.

On the one hand the membrane is considered the heart of a redox flow battery. On the other hand, the bipolar plate is one of the key components of an RFB. However, the Bipolar plate is important, since the plate has an impact on the complete systems, as far as total dimensions, total weight, thermal and electrical properties of the stack and thus of the system is determined by the bipolar plate technology [2].

As already mentioned, the chemical conditions for the materials used in redox flow batteries are challenging [3]. Most systems are operated between 40°C and 60°C in a liquid of dissolved vanadium salts in sulfuric acid. Besides the Vanadium-technology, there also some other technologies (metallobased or organic RFBs), which will not be further considered.

Due to these harsh conditions, superimposed by an electrochemical potential, graphite-based bipolar plates with polymeric binders are used in almost all applications in these battery stacks. The graphite composite plates are an unbeatable material in terms of stability under the above-mentioned corrosive conditions, and the cost-intensive coated metal plates have no chance.

They have been operated very adequately several times over the years. However, due to an intrinsic fragility caused by a high filling load with graphite, graphite composite plates require a greater thickness than metal plates, resulting in more weight and volume of the stack. From a cost point of view, the membrane is certainly considered the dominant part of the redox flow battery stack. However, the bipolar plates tend to be underestimated both in terms of their technical requirements and, in particular, their contribution to the cost structure.

Graphite composite based bipolar plates are manufactured using highly filled compounds [2]. They contain fillers like graphite and/or other electrically conductive carbons incorporated in polymers performing as a gluing binding matrix. The key challenge is the competing interaction between electrical conductivity - achieved by the carbon component - and mechanical stability as well as liquid tightness which is provided by the binding polymer.

The compounding process is the first step to produce highly filled, electrically and thermally conductive pellets for the subsequently following step of forming bipolar plates.

Both compounding and molding processes, which can be injection molding, compression molding or continuously extrusion, are very sensitive to process parameters and need to be carefully controlled. The objective is to manufacture bipolar plates in large volumes and high quality more or less like standard plastic parts. Only by using price cost attractive materials and the consequent focus on process automation by higher volume, the bipolar plate can contribute significantly to a better market acceptance of RFB.

Besides the bipolar plate, the gasket is a very important component of the battery stack and tends to be heavily underestimated. It plays a key role in the mechanical properties of the stack. Inappropriately selected gasket materials may cause cracks in the bipolar plates or may affect the membrane-structure negatively. Despite the fact that the gasket has to seal the stack, the cooperation with other stack components and their cumulative tolerance effects have to be on focus for the stack design and for the operation of its.

The same which is evident for each component is also obviously for the gasket; they have to be cost attractive. Therefore, in some research projects, it is the objective is to suspend gaskets completely and use welding or bonding processes instead.

Technically, the bipolar plate of a RFB stack has to accomplish the following functions [3, 4]:

- conduct electrical current,
- conduct heat and distribute coolant in a eventually incorporated cooling flow field,
- provide mechanical stability of the stack,
- prevent permeation and leakage

However, the functions of the gasket are completely different. The main functions of gaskets in a RFB stack are [5, 6]:

- sealing and leakage prevention of anode and cathode area,
- sealing and leakage prevention of cooling plates,
- compensate tolerances and dimensional changes during stack-assembling caused by interaction with all stack components.

2. Technical requirements of bipolar plates and gaskets

Based on the technical functions described above, a comparison to other technologies is necessary: The Fuel Cells: The US department of energy (DoE) suggested development targets for fuel cell components as shown in the **Table 1** for bipolar plates [8]. Although these data are based on communication and data from

Technical property	Units	Targeted value
Plate weight	$\frac{kg}{kW}$	< 0.4
Electrical conductivity Depending on type	$\frac{S}{cm}$	> 100
Thermal conductivity	$\frac{W}{m \cdot K}$	> 10
Flexural strength	MPa	> 34
Shore D hardness		> 40
Temperature resistance Thermo-mechanical test	°C	> 70
Acid uptake Depends on application or technology		low

Table 1.
Benchmarks for bipolar plates in redox-flow applications defined by DoE [7] and experiences from customer requirements from Eisenhuth GmbH & Co. KG.

Technical property	Units	Targeted value
Density	$\frac{g}{cm^3}$	1,15 to 1,5
Electrical conductivity Depending on type	$\frac{S}{cm}$	< 1
Shore D hardness		< 70 , preferred value ~ 40
Compression set	%	< 18
Temperature resistance Thermo-mechanical test	°C	> 70
Chemical stability Depends on RFB-type	Resistant against the used chemical environment; no or low changes in properties (typically mechanical)	

Table 2.
Proposed benchmarks for gaskets in RFB based on fuel cell requirements [9] and experiences from customer requirements from Eisenhuth GmbH & Co. KG.

conventional low temperature PEM fuel cell developers, most of the targeted values can be directly transferred to Redox-Flow technology.

Additionally, the chemical resistance of the bipolar plate can be characterized by measuring the corrosion current under a potential typical for RFB and using sulfuric acid or something suitable (depending on application as an electrolyte). The detailed parameters and development objectives of this corrosion test are still subject to technical discussions and depend on the anticipated application of the plate. A similar table of functional requirements can be set-up also for gasket materials in RFB.

The gasket material has to be resistant against the selected electrolyte and environment under operating conditions. This is qualified for example by comparison the mechanical properties of recently produced and altered samples. It has to be noted that the values mentioned in **Table 2** are for orientation and refer to standard elastomer materials available on the market. Based on these technical requirements, an appropriate feedstock respectively materials for both bipolar plates and gaskets have to be selected.

3. General concepts of bipolar plate manufacturing

As mentioned above, composite bipolar plates consist of a binder polymer, which is highly filled with a conductive carbon component. Typical compositions are >80 wt.% conductive filler and < 20 wt.% binder polymer. Compounding, processing and manufacturing is substantially different from conventional polymers due to the high content of filler material in the compound [10]. The function of the carbon filler is to provide electrical and thermal conductivity.

Therefore, a three-dimensional percolating carbon structure is required. Usually, the main carbon component of the plate is synthetic graphite and the second material is carbon black. For producing plates, several options are possible:

- Compression molding
- Injection molding
- Plate Extrusion
- Foil Extrusion

In all methods, after removal from the process certain after treatment procedures may be necessary. Either to remove the 'skin' of the mold release agent from the surface of the plate or as noted in Derieth et al. [9, 10] to remove an accumulation of polymer from the injection. Or compression molding skin.

3.1 Binder polymers

In general, two different concepts of polymer binders can be applied in bipolar plates. First, the binder material can be polymerized or cross-linked in-situ in the composite during molding of the plate (resin method). The used polymer is thermosetting, which provides good mechanical properties at elevated temperatures and often a relatively easy processing [11].

Second, a thermoplastic polymer material can be used in the compounding process (thermoplast method). Since the most materials in RFB are thermoplastic materials, in the following the focus will be also in this consideration. The polymer has to be selected with sufficient chemical, mechanical and thermal stability (e.g. data from [11]). Several material candidates are available on the market in high quality and well-defined configurations for different processing methods and applications due to the usage of additives like waxes, minerals or fibers.

Figure 1 shows the well know pyramidal classification for more than a handful of popular plastics.

3.2 Graphite materials and fillers

While graphite is generally the major filler for bipolar plates to achieve a sufficient conductivity, several other carbon additives can be employed in order to boost conductivity properties of the composite material. Examples for such additives are highly conductive carbon nano tubes (CNT), high surface carbon blacks (CB) or multi-layer graphene nanoplatelets [7].

Due to its crystalline layer structure graphite is inherently anisotropic in its physical properties e.g. electrical conductivity or its mechanical behavior. Electrical conductivity is being provided by the mobility of electrons within the graphite layers of each platelet. Contrary to the conductivity along the layers, graphite is perpendicular to these layers an electrical insulator. Thus, the bipolar plate manufacturing process should 'promote' different orientations of the platelets forming isotropic physical properties of the macroscopic plate material. Some additives such as carbon blacks are helpful to increase the number of conductive paths in the carbon-polymer-system. The nano-sized carbon blacks do function as a 'gap-filler' in the insulating polymeric matrix between the micro-sized graphite particles and this in consequence increases the overall material conductivity significantly [7].

The overall conductivity in a bipolar plate is generated by a three-dimensional percolating network which consists of conductive particles. The carbon-binder system is always inhomogeneous and can be considered as a two-phase system of conductive carbon paths bonded in a polymer matrix as shown in **Figure 2**. The structure of the material highly depends on the chemical composition and not less

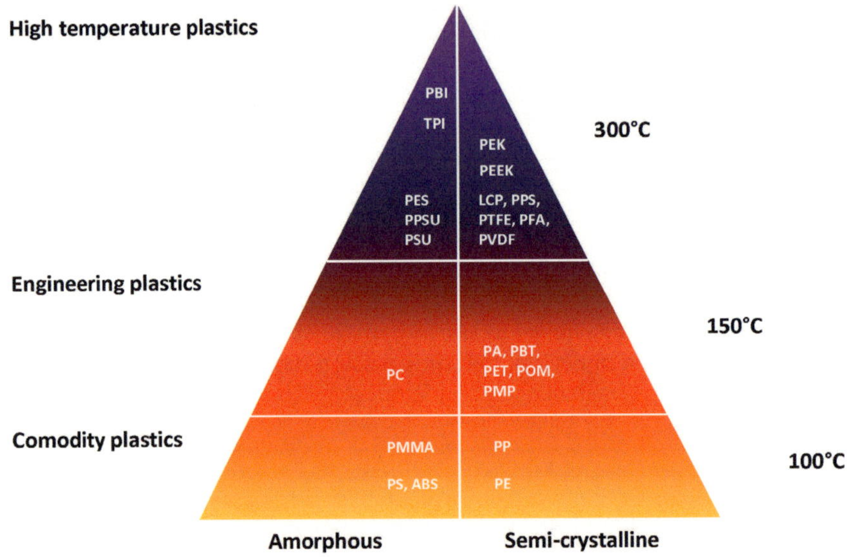

Figure 1.
The plastics pyramid preferred materials for RFB applications are semi-crystalline materials such as PP, PE and PVDF [9].

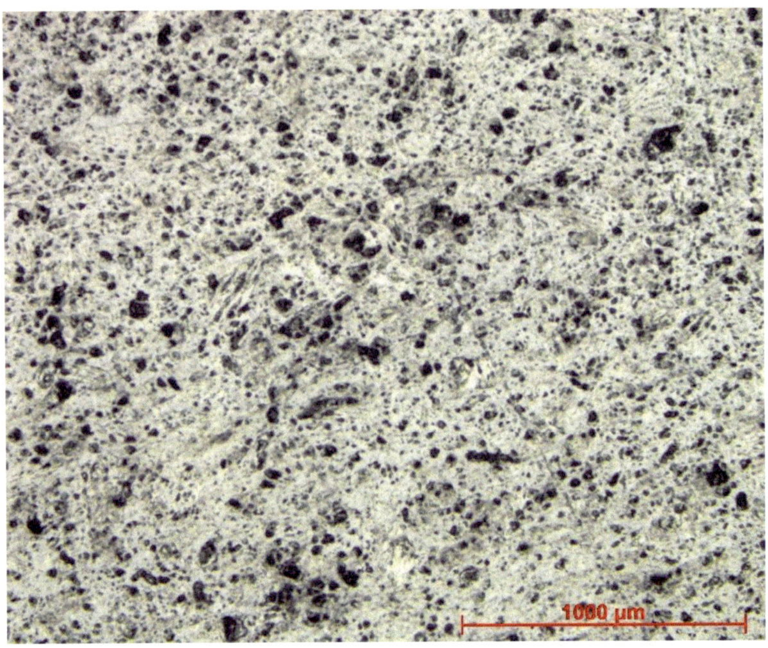

Figure 2.
Polarization microscopy of polished surface from bipolar plate with 80 wt.% graphite content. Particles (black) are locatable in polymer matrix (white).

important on the kind of the chosen processing-approach (compounding, molding, extruding...) and the therefore used parameters. The complete processing chain – from the raw material to the molded plate – has to be carefully controlled to ensure consistency and reproducibility of the bipolar plates.

Carbon blacks can be formed in the gas phase by thermal decomposition of hydrocarbons under different conditions [11] and this results in a broad variety of materials with differences in surface area, hydrophobicity and conductivity. The different carbon black grades are then available for the adequate application and function.

Keeping this in mind it has to be considered that high surface carbons are more disposed to (undesired) carbon corrosion effects than graphite-based materials, thus their positive conductivity effect has to be balanced against long term stability requirements.

Another important aspect of carbon materials is purity. Since most fuel cell membranes and catalysts are highly sensitive against contamination with Iron-ions and other metal residuals, the raw materials for bipolar plates have to be carefully characterized with respect to their contamination level. The carbon or graphite type also mainly determines bipolar plate's properties like porosity, phosphoric acid uptake or corrosion and hydrophobicity, both regarding the surface and the bulk [11, 12].

3.3 Recycling

At status quo, the amount of waste caused by the production of bipolar plates – an inhomogeneous system consisting of plastic and carbon –is significantly higher when compared to a fully implemented commodity plastic production process [13]. The waste accrues in form of rejects from production, which can be lowered by optimization processes, but also in form of gates, which are necessary for production and dimensioned by material properties. In addition, the systems in which the material is used have a limited lifespan, so the demand of reusage of the parts made of graphite compound or the compound itself is conceivable.

On the other hand, there is the possibility to use secondary materials as feedstock of the graphite compounds to substitute fully or partially the conventional fillers. Conductive fillers like synthetic graphite are valuable resources being produced via different thermal processes, which are similar to other thermal processes e.g. some recycling processes for various other wastes. Some of these processes generate in some degree useful carbon materials [13].

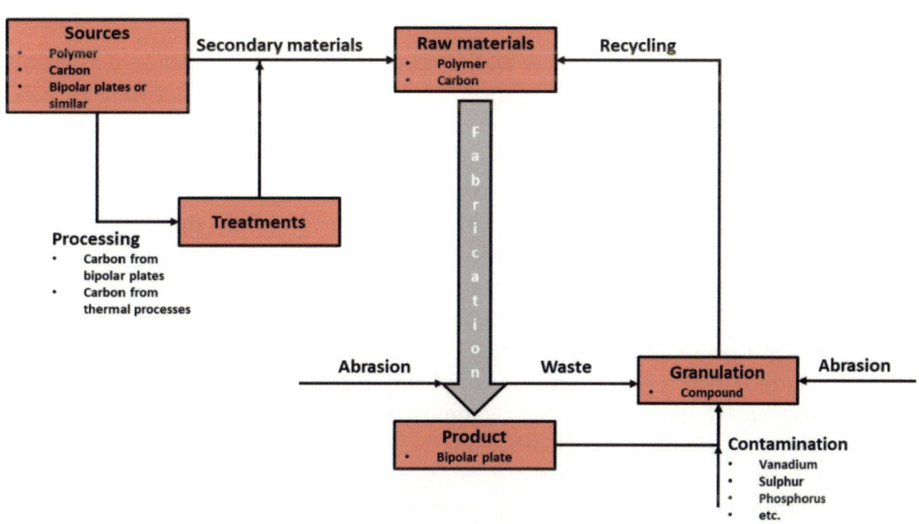

Figure 3.
Scratch of flow diagram for resources from recycling and secondary sources. The primal structure is from plastic treatment [14–16]. The obvious barriers are the contaminations and changes in material properties caused by multiplied processing.

These circumstances and opportunities result in an increasing development of recycling methods with the consequence of a property upgrade of the carbon by combining lower general production costs. In the best case these carbons are suitable for bipolar plates. In **Figure 3** the principle of the different recycling opportunities are being described.

4. Characterization data of RFB bipolar plate materials

Certainly, the final criteria of success for any bipolar plate is the in-situ control of performance and stability under real RFB conditions. However, RFB are highly complex systems with numerous sources of inconsistency. Thus, standardized ex-situ bipolar plate characterization is required for material development and quality control. Several test methods are well established for bipolar plates and some are presented below. The list of test methods is not considered to be complete.

4.1 Electrical conductivity measurements (In-plane)

Clearly, electrical conductivity both in-plane and through-plane is one of the most important properties of the bipolar plate. Despite most fuel cell (component) and battery laboratories have access to electrical conductivity testing equipment, by now there is no generally standardized test method for bipolar plates for RFB, and comparing results from different sources can show significant differences, even though the same samples are tested. One of the main reasons may be surface effects and pre-treatment of the sample. As shown in the **Figure 4**, Eisenhuth has implemented a testing system for this application, which is suited for local in-plane conductivity testing with the option to measure several times at different locations on the plate.

The in-plane conductivity device allows for a conductivity mapping over a sample area of 750x300 mm. Thus, the characterizing of the plates with respect to the degree of homogeneity during production is possible. Conductivity mapping is an important tool both for quality control as well as and furthermore for the material and process development.

For graphite composite plates it is well known that compounding and molding are highly sensitive to process details and may generate inhomogeneous structures on the surface and/or in the inner core of the material. Certainly, the development target is a homogeneous distribution of conductivity with only minimal deviations between different points on the bipolar plate.

For PPG86 and BMA5 or BMA6 plates the compounding and manufacturing process in hot pressing are established and well controlled, and the conductivity mapping shows an even distribution. Irregularities in the conductivity are in some processes unavoidable because of the process-depending-orientation of the particles through different processing influences. For example, in injection molding the filler particles orientate differently from the core to the surface of the produced parts, which results in different conductivities measured In-plane or through-plane. In addition, the regions which will be filled lastly in injection molding show a higher average conductivity compared to the gate region.

A conductivity mapping of a second process example for a PPG86 based plate is shown in **Figure 5**. This specific plate is produced by plate extrusion. The border area of the plate parallel to the flow direction during the extrusion process seem less conductive.

In terms of hot pressing – a process with a certainly low flow – these described irregularities are more dependent from the overall process stability and experience

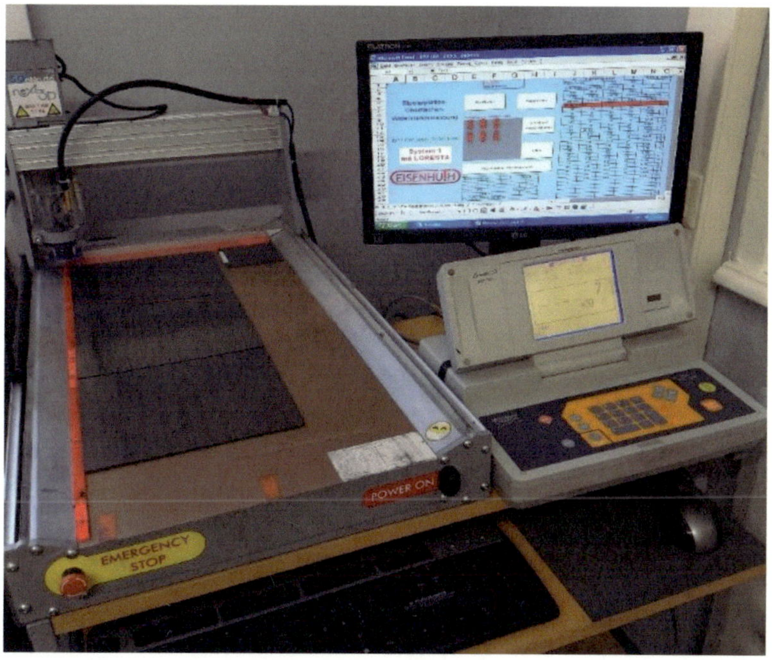

Figure 4.
Testing device and method for electrical conductivity measurement at Eisenhuth GmbH & Co. KG.

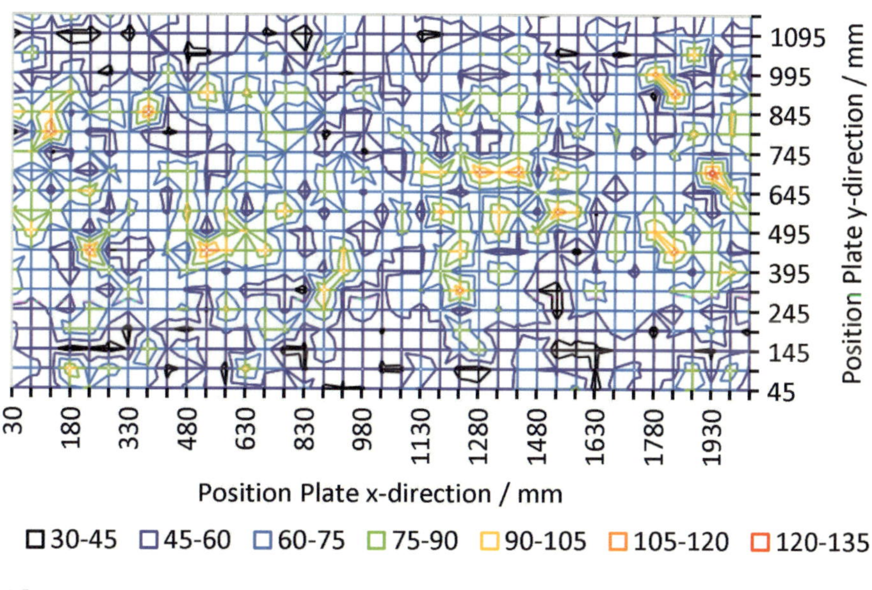

Figure 5.
In-plane conductivity mapping of a PPG86 based bipolar plate made by extrusion by another company who is also active in the field of redox flow batteries. The material shows a higher resistivity at the outside areas caused by the manufacturing-process. The results are corrected with finite size corrections for 4-point probe measurements [17].

of the manufacturer. Development in the field of hot pressing by Eisenhuth GmbH & Co. KG in the last years are focused mainly in material research with the aim of reaching higher conductivity, larger plate designs and simultaneously easier production.

In **Table 3** technical properties for bipolar plates made by hot pressing are shown. The data is measured with the shown in-plane conductivity measurement

Technical property values	2018		2020	
	PPG86	BMA6	PPG86	BMA6
Density ($\frac{g}{cm^3}$)	1.8	2.1	1.8	2.0
Electrical conductivity ($\frac{S}{cm}$) Depending on type	96.2	192.3	185.2	312.5
Flexural strength (MPa)	21.1	31.8	22.4	31.7

Table 3.
Technical properties from databases from Eisenhuth GmbH & Co. KG.

device and specimen according DIN EN ISO 527 tested on a universal testing device and a microbalance.

The comparison between the results shows that the improvement of the standard products from Eisenhuth GmbH & Co. KG has led to an increase of the electrical conductivity from around 75%. But the mechanical behavior seems similar. This is due to optimizing process parameters and periodic testing of new raw materials.

4.2 Qualification of secondary raw materials by thermogravimetric analysis

As described above the material used for the bipolar plates in RFB applications is made out of plastics and conductive fillers like graphite. During RFB operation the bipolar plates are exposed to normal temperatures, such as 40°C. Consequently, all raw materials used for plate manufacturing have to resist approximately 40°C.

Parallel to the shortage of the raw materials, the Vanadium-RFB technology has to compete regarding cost- and technology-aspects to other technologies, in particular with the lithium ion storage technology. Knowing this background, it is more than advisably to look out for alternative materials. Thus, the Eisenhuth GmbH & Co. KG is investigating together with a consortium alternative material sources, in particular from the recycling sector. Two potential processes which produces carbon materials are shown in **Figure 6.**

Both processes separate carbon in form of agglomerated particles. Tyres consists of rubber filled with carbon pigments to strengthen the material. By the oxygen-less combustion of tyres the carbon will be released and during the pyrolysis process it is being formed to agglomerates.

During methane cracking – a process to produce hydrogen - carbon can agglomerate on particles, which function as. The particles consist of contaminations of the used feed gas or are part of the used catalyst [17].

The samples are called CB-RC for the tyre recycling carbon black and CB-MC for the methane cracking carbon black. Resulting curves of the mass loss over the temperature of TGA from different carbon blacks are shown in **Figure** 7. Samples of conventional carbon blacks are called CB-C.

The thermogravimetric analysis (TGA) can be used in order to determine the combustion and vaporization temperatures of the materials and allows to quantify the contents of different materials in the compound [19].

During TGA a sample is heated under defined conditions such as gas environment and heating rate. The weight loss as well as the temperature (in correlation to the weight loss) of the sample in the oven is being determined. The TGA is used at Eisenhuth as an instrument of permanent quality control of the process. It also can be used, to get more information about the compound material.

The TGA curves show that the secondary materials contain a high content of ash. The influence of the ash is at that point unknown. At best it does not influence

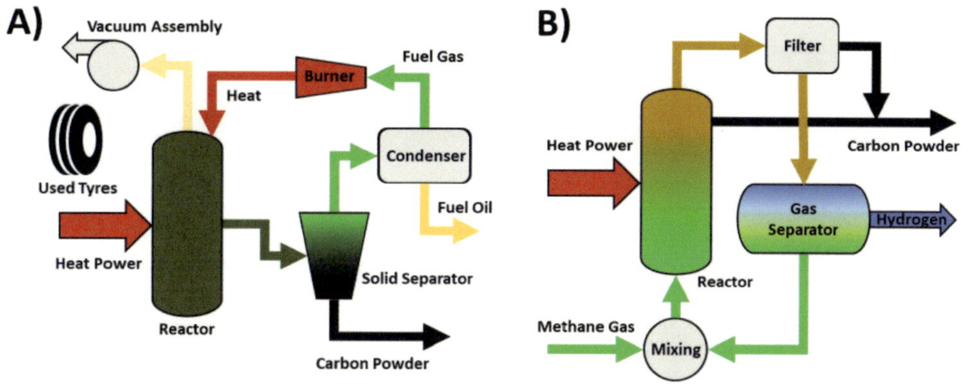

Figure 6.
Principle of producing carbon black and graphite from used tires (A) and in a hydrogen production in a methane cracking process (B) [13, 16, 18].

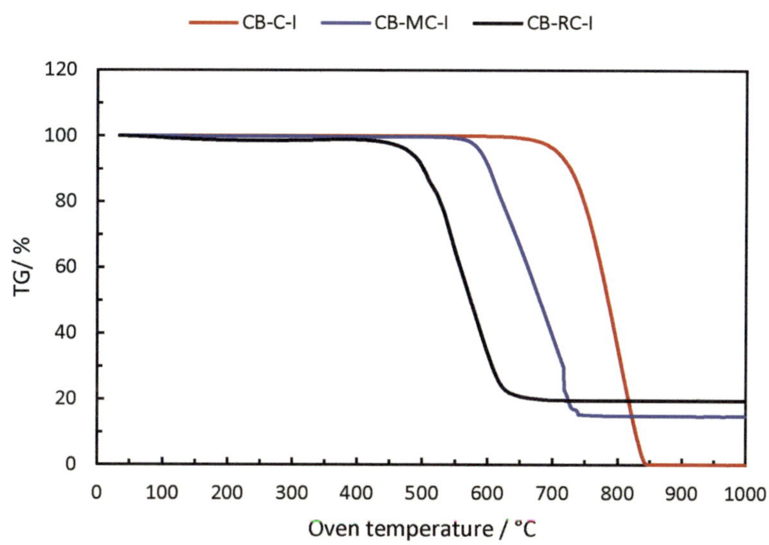

Figure 7.
TGA data from different carbon blacks. Conventional CB (CB-C) and CB from secondary sources (CB-MC and CB-RC) are compared. The differences in combustion temperatures and contents of ash are significant.

Carbon black type	Average combustion temperature / °C	Rest mass (Ash) / %
CB-C-I	784.4	0.13
CB-C-II	691.4	−0.29
CB-MC-I	684.7	8.66
CB-MC-II	666.4	14.86
CB-MC-III	695.8	10.18
CB-RC-I	558.4	19.52
CB-RC-II	541.4	12.45

Table 4.
Results of TGA from different CB-types. The CB tested are conventional (CB-C), products from methane cracking (CB-MC) and products from Tyre recycling (CB-RC). The average combustion temperature is at the point of 50% weight loss of the combustible mass.

CB used in compound	Filler content / $wt.\%$	Compound conductivity / $\frac{S}{cm}$	Reference conductivity / $\frac{S}{cm}$	CB used in Ref.
CB-C-I	78	28.3 ± 2.0	13.2 ± 2.5	None
CB-C-II	75	12.9 ± 0.9	< 1	CB-C-I
CB-MC-I	80	11.3 ± 1.2	< 1	None
CB-MC-II	80	10.6 ± 2.5	< 1	None
CB-MC-III	80	12.0 ± 1.5	< 1	None
CB-RC-I	75	8.9 ± 0.7	18.7 ± 0.6	CB-C-I
CB-RC-II	75	9.0 ± 0.8	18.7 ± 0.6	CB-C-I

Table 5.
Results of conductivity measurements from different compounds. Partially consisting of the described CB types from conventional and secondary sources compared to individual references produced parallel the tested compound mixtures.

or at least it has a minor influence on the properties of the compound respectively the plates. In the worst case some critical contaminations are soluble or volatile and will damage the system, in which the material is used. The details are shown in **Table 4**. The average combustion temperature is extracted from the curves at the point at which 50% of the weight of the combustible mass is lost.

The second that stand out is the difference of the combustion temperatures of the tested CB. A lower combustion temperature under the assumption that the tested materials consists of similar carbon structures is an indication for a higher surface area [20]. It is described that the surface area – normally measured for CB according ASTM D 2414 with dibutyl phthalate (DBP) – has an influence in the percolation threshold and the resulting conductivity of the corresponding compound. The percolation threshold is the small zone in which the compound receives a mayor increase in its electrical conductivity by only adding a very less of filler content [21].

In order to characterize the influence of the different CB types on the conductivity, the secondary CB are integrated and evaluated in various testing and production series to compare the new materials with the current neat carbon black. The **Table 5** shows the results of conductivity measurements like described above from different compounds, in which the CB types are used. For comparison individual references from the mentioned testing and production series are listed in the same table. The compounds are made by combining different polymers mostly PP with graphite. Some of the graphite is replaced with the different CB to keep the recorded filler content at the same level for all.

It can be observed that the impact of the secondary CB on the electrical conductivity is noticeable, but far less for CB-MC and CB-RC-types than the qualification by the TGA suggests. The compounds consisting the CB-MC types have a relatively low conductivity compared to standard materials but the reference compound with the same filler content has no measurable conductivity, therefore the CB-MC types seem to reduce the percolation threshold for the filler in the compound. The CB-RC types have compared to conventional CB a smaller impact on the conductivity because half the value of the CB-C-I consisting reference compound with the filler content of 75 wt.% has been measured.

The qualification by the TGA was fitting for the conventional types. Whereas the high differences between the prognosis and the measurement results for the secondary CB types are unexpected and a high level of uncertainty remains. The reason of these differences can be the high content of probably non-conductive contamination or different carbon structures of the particles. Both reasons are possibly responsible for a way lesser combustion temperature during TGA-measurements. The higher the lever of amorphous carbon and impurities so lower the combustion temperature and the achievable level of conductivity.

5. Gaskets application and qualification

Since many years the fuel cell developers invested tremendous efforts in improvement and technological readiness of the core components, such as membranes and electrodes configuration. However, within the last years the gasket material was recognized more and more as an underestimated component. Despite the gasket does not directly contribute to the electrochemical processes, inappropriate gaskets can cause leakages [6].

The increased use and establishment of the systems on the market, primarily among consumers, has resulted in a focus on safety issues during consumption and error sources during mass production.

Common hard gaskets support well defined gaps, however may be compromised in their sealing properties, do not compensate tolerances very well and may put mechanical distortion on the bipolar plates, which can cause cracks or breaking after a long time. On the other hand, with soft gaskets it is more difficult to control the performance of the system cause of limitations in parameters like pressure. These descriptions are analogue to RFB systems.

In general, like the other RFB components the gaskets have to resist temperatures up to 70°C, electrolytes like sulfuric acid or other materials of RFB systems like bromine and contact to electricity. Fluoroelastomers (FKM) and ethylene propylene diene monomer rubber (EPDM) are most likely the materials of choice for several applications.

For certain applications EPDM might be a cost-efficient alternative for systems which can handle the stiffness of this rubbers. The arguments clearly show that gaskets are a highly customized component for each stack manufacturer. For overview, some typical gasket properties for a broad variety of materials are shown in the **Table 6**.

Along with the rubber materials like silicone (SI), hydrogenated nitrile butadiene rubber (HNBR), EPDM and FKM thermoplastic elastomers (TPE) in form of styrenic block copolymers (TPS), thermoplastic vulcanizates (TPV), thermoplastic polyurethanes (TPU) and thermoplastic polyolefin elastomers (TPO) are listed. These thermoplastic-elastomers have similar properties to rubber but can be processed like "common" thermoplastics and can be softer if required. This has the advantage of easier manufacturing and recycling of the material as well as a broader range for applications.

5.1 Qualification of physical properties affecting gasket production

In order to supply consistent gaskets with appropriate tolerances the viscoelastic properties of the gasket prepolymers and thermoplastics are an important parameter. A low viscosity is beneficial for processing. For plastic materials usually the mold flow rate or mold flow index are specified by the supplier, supporting the manufacturer for plastic parts with processing-relevant- information and –parameters. However, these data are 'standard data' and not always compatible with the molding conditions or equipment at the part manufacturer.

In addition, for rubber materials or their pre-polymers and thermoplastic elastomers these data are not available in most cases, because of their impacting

Description and Unit	TPS	TPV	TPU	TPO	EPDM	SI	FKM	HNBR
Hardness Shore A	2–95	20–95	2–85	65–95	25–85	25–85	50–90	40–90
Temperature range °C	−50 / +120	−40 / +130	−40 / +85	−40 / +70	−50 / +100	−70 / +200	−20 / +220	−30 / +150
Steam resistance	—	—	—	—	++	++	++	++
Oil resistance	+	+	—	+	—	+	++	++
Acid /bases resistance	++	++	0	++	++	—	++	+

Table 6.
Gasket material overview with typical physical properties and behavior in system specific conditions used by Eisenhuth GmbH & Co. KG.

viscoelastic behavior. Therefore, Eisenhuth developed a phenomenological test method to characterize polymer materials with respect to processability. In this test, a melt of the used pre-polymer or thermoplastic is pressed into a spiral-shaped mold with a defined pressure under process-relevant temperatures.

The viscous melt flows into the spiral mold and finally stops, when the applied pressure is equal to the 'back-pressure' in the mold. The reason therefore is the higher lever of the progressing polymerization, vulcanization or solidification of the melt. The length of the helix can be correlated to the viscosity and consequently to the processability. The longer the helix the lower the viscosity. This is helpful to find the processing ranges of materials as far as the viscosity is concerned (**Figure 8**).

As mentioned, the length of the spiral is an good indicator for the processability. This test has been performed with a variety of potential gasket materials to achieve a data baseline. The values are shown in **Figure 9**.

The results show that processability depends strongly on the material but different types of the same material have also high differences. Exemplarily shown in the Figure the good processability of some types of thermoplastic elastomer materials cannot be reached by the measured processability of rubber materials.

5.2 Chemical resistance of gasket materials in system specific environments

The processability of the thermoplastic elastomers is convenient but it is necessary to qualify the mechanical and chemical stability of the materials. The called rubbers are commonly used in different applications such as fuel cells and chemical industries, and their long-life behavior is already known.

To ensure the stability of the materials specimen according DIN EN ISO 37 are made and treated in this example in vanadium electrolyte for VRFB applications.

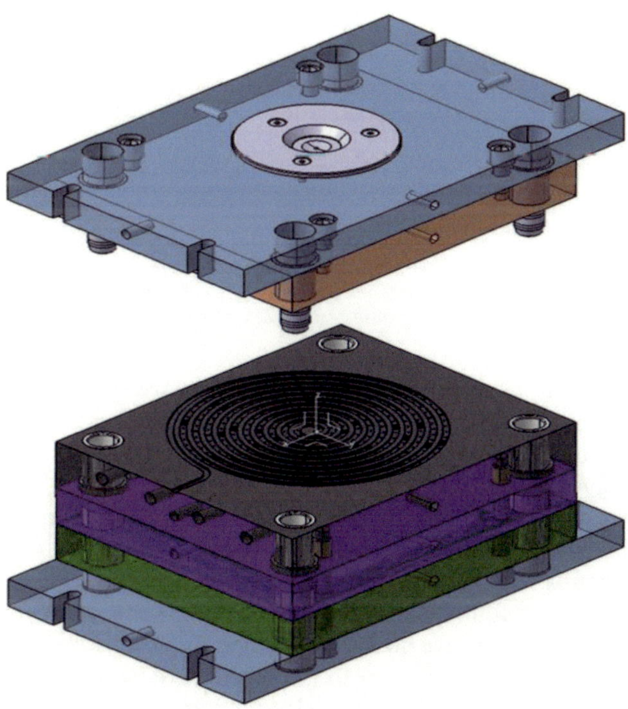

Figure 8.
Spiral mold for characterization processability by injection molding by Eisenhuth GmbH & Co. KG.

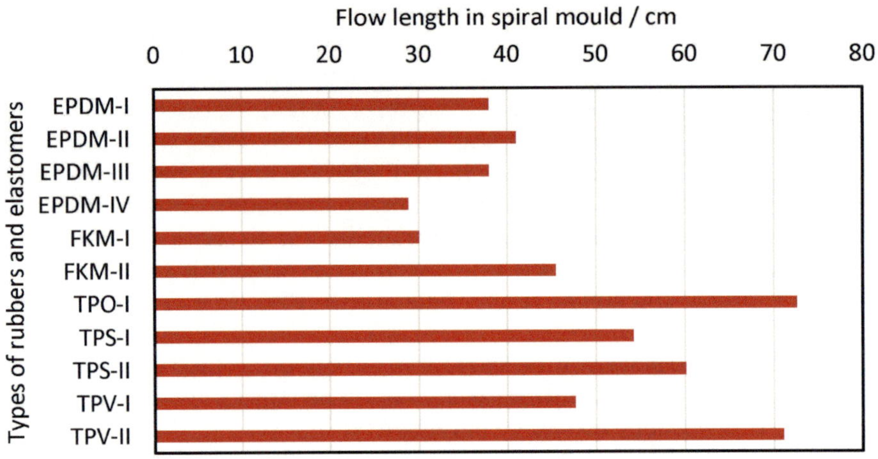

Figure 9.
Results of injection molding in spiral mold.

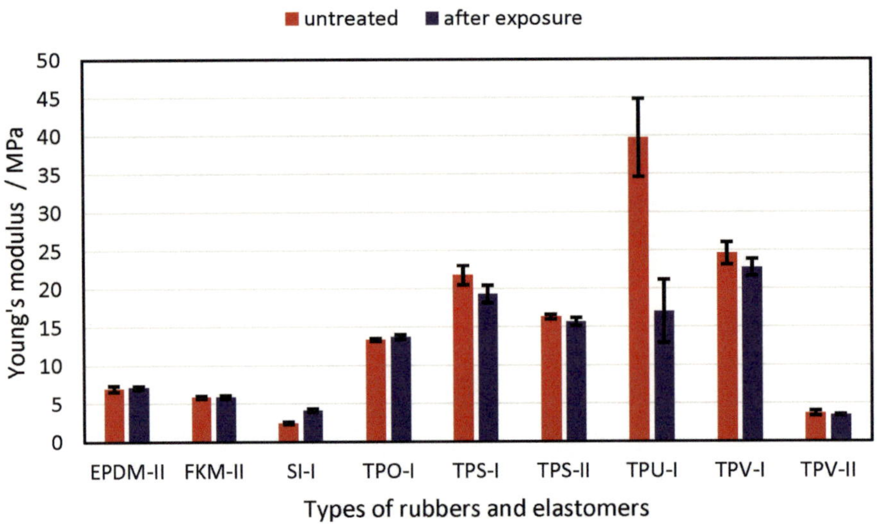

Figure 10.
Results of tensile strength testing according DIN EN ISO 37. The tested materials are different rubbers and thermoplastic elastomers. The untreated specimens are directly tested after production. The second data is generated after exposure of specimens with aggressive vanadium electrolyte used in VRFB.

The electrolyte is positive charged, so the most aggressive species of vanadium ions is to 1.65 mol/L concentrated in 4 mol/L sulfuric acid. The specimens are treated the same time for around thousand hours and therefore stored in glasses with full surrounding electrolyte. The specimens are tested for tensile strength according DIN EN ISO 37 in a universal testing machine. The resulting Young's modulus are shown in **Figure 10** and are exemplary for the overall changes in mechanical properties of the treated specimens.

It is shown that the stability of the rubbers in the specific environment is good. The modulus is low but there are no major changes measured. In average the thermoplastic elastomers are different to the rubbers. Most of the materials and material types have a higher modulus with low changes. TPU seems not suited for the application, moreover one indication are high changes such as superficial cracks in the surface of the sealing.

6. Conclusions

In this chapter the basics and advantages of graphite bipolar plates could be presented in connection with current research topics at Eisenhuth regarding the reduction of production costs and the related easier market introduction of RFB. Furthermore the suitability of easy to process thermoplastic elastomers as sealing material in RFB was shown.

It was explained that the proposed targets for material properties are not fully achieved, but that progress in materials research is possible. For example, the electrical conductivity of standard materials for RFB could be optimized by about 75% in recent years.

Options to reduce costs through recycling methods and use of secondary resources were discussed. It could be shown that the substitution of commercial carbon types such as synthetic graphite by secondary materials for composite production is possible.

The differences in the processability of rubber types and thermoplastic elastomers were shown by tests in a correspondingly designed injection mold. The chemical stability of some types of thermoplastic elastomers is tested for VRFB.

Acknowledgements

The authors acknowledge fruitful collaboration, extensive test work and the positive relationship to Technical University of Clausthal, German Aerospace Center Oldenburg (Institute of Networked Energy Systems) and Research Center of Fuel Cell Technology (ZBT) in Duisburg.

Public funding is gratefully acknowledged from Federal Ministry for Economic Affairs and Energy (Germany) and Ministry of Education and Research (Germany) in cooperation with the project holder Forschungszentrum Jülich in the projects 'Redox Flow Extrusion', 'Re3dox', 'Demo-Bio BZ' and from the State of Lower Saxony in the project "Maleskues" and "Titan Porous Hybrid".

Conflict of interest

The authors are part of the company Eisenhuth GmbH & Co. KG, which produces bipolar plates made of graphitic compounds and gaskets for fuel cell, redox-flow battery and heat exchanger purposes.

The shown data are part of the acknowledged public funded projects. The conclusions and statements made are based on the experience of the authors in their specific working fields in the said company.

Author details

Thorsten Hickmann*, Toni Adamek, Oliver Zielinski and Thorsten Derieth
Eisenhuth GmbH & Co. KG, Osterode am Harz, Germany

*Address all correspondence to: t.hickmann@eisenhuth.de

References

[1] Renewable Power Generation Costs in 2018, International Renewable Energy Agency, Abu Dhabi, 2019. https://www.irena.org/-media/Files/IRENA/Agency/Publication/2019/May/IRENA_Renewable-Power-Generations-Costs-in-2018.pdf (accessed November 28, 2019).

[2] Apelt S, Hickmann T, Marek A, Widdecke H. How conductive compounds work. Kunststoffe International, No 12. 2006. 86-90 p

[3] J. Noack, N. Roznyatovskaya, T. Herr, P. Fischer, The Chemistry of Redox-Flow Batter-ies, Angew. Chem. Int. Ed. Engl. 54 (2015) 9776-9809. https://doi.org/10.1002/anie.201410823.

[4] Larminie J, Dicks A, editors. Fuel Cell systems explained, first edition. New York: Wiley. 2000. DOI: 10.1002/9781118706992

[5] Kakati B Kr, Verma A, editors. Carbon polymer composite bipolar plate for PEM fuel cell. Saarbrücken: LAP LAMBERT Academic Publishing. 2011. ASIN: B01K955QYA

[6] George M. Vanadium redox flow batteries: design and experimentation. Honors Research Projects, 666. 2018. Available from: http://ideaexchange.uakron.edu/honors_research_projects/666

[7] Hickmann T. Plastic applications in PEM fuel cells. VDI Reports, No 2035. 2008. 81-83 p

[8] Office of Energy Efficiency & Renewable Energy [Internet]. Available from: https://www.energy.gov/eere/fuelcells/doe-technical-targets-polymer-electrolyte-membrane-fuel-cell-components, [Accessed: 2020-08-31]

[9] Bonnet M, editors. Kunststofftechnik. Hamburg: Springer Verlag. 2013. ISBN: 978-3-658-03139-8

[10] T. Derieth, G. Bandlamudi1, P. Beckhaus1, et al. Development of Highly Filled Graphite Compounds as Bipolar Plate Materials for Low and High Temperature PEM Fuel Cells, Journal of New Materials for Electrochemical Systems 11, 21-29 (2008)

[11] Pierson H, editors. Handbook of Carbon, Graphite, Diamond and Fullerenes. USA: Park Ridge [USA]: William Andrew. 1993. ISBN: 9780815513391

[12] Huijun L, Lingxu Y, Qian X, Chuanwei Y. Corrosion behavior of a bipolar plate of carbon–polythene composite in a vanadium redox flow battery. RSC Adv., Volume 5. 2015. 5928-5932p. DOI: 10.1039/c4ra13697g

[13] Hickmann T, Zielinski, O.: Redox Flow Battery: System for Test Series with Recycling Material in: Conference proceedings: ICEES 2020 - 4th International Conference on Energy and Environmental Science, Pert h, Australia, January 8-10, 2020

[14] Stübler N, Hickmann T, Ziegmann G. Effect of methanol absorption on properties of polymer composite bipolar plates. Journal of power sources, Volume 229. 2013. 223-228 p. DOI: 10.1016/j.jpowsour.2012.11.129

[15] Jörg Woidasky Marc-Andree Wolf. In: Peter Elsner Peter Eyerer Thomas Hirth. Kunststoffe, Eigenschaften und Anwendungen. Heidelberg: Springer-Verlag. 2012. 105-113 p. DOI: 10.1007/978-3-642-16173-5

[16] Schroder D. Semiconductor Material and Device Characterization. Hoboken: Wiley. 2006. 1-59 p. ISBN: 978-0-471-73906-7

[17] Groves I, Lever T, Hawkins N. Determination of Carbon Black Pigment

in Nylon 66 by TGA. TA Instruments Ltd. (U.K.) [Internet]. Available from: http://www.tainst.com

[18] Roy C, Labrecque B, de Caumia B. Recycling of Scrap Tires to Oil and Carbon Black by Vacuum Pyrolysis. Resources Conservation and Recycling, Volume 4. 1990. 203-213 p. DOI: 10.1016/0921-3449(90)90002-L

[19] Abánades A. Low Carbon Production of Hydrogen by Methane Decarbonization. In: Zhen F, Smith R, Xinhua Q, editors. Production of Hydrogen from Renewable Resources. Netherlands: Springer-Verlag. 2015. 149-177 p. DOI: 10.1007/978-94-017-7330-0

[20] METTLER TOLEDO. Analysis of Elastomers with Different Types of Carbon Black. METTLER TOLEDO TA Application Handbook Elastomers, Volume 2 [Internet]. Available from: https://www.mt.com/de/de/home/supportive_content /matchar_apps/MatChar_HH456.html [2020-08-31]

[21] Xiaoyu L, Hua D, Qin Z, Feng C, Qiang F. The effect of DBP of carbon black on the dynamic self-assembly in a polymer melt. RSC Adv., Vol 6. 2016. 24843-24852 p. DOI: 10.1039/c5ra28118k

Chapter 3

Hybrid Nature Properties of $Tl_{10-x}ATe_6$ (A = Pb and Sn) Used as Batteries in Chalcogenide System

Waqas Muhammad Khan and Wiqar Hussain Shah

Abstract

In future, the most common batteries will be the thallium. As there is many types of batteries but the thallium batteries are better from them. In here, we have made the compound which is more positive work than the other batteries. The different elements are doping in the tellurium telluride to determine the different properties like electrical and thermal properties of nanoparticles. The chalcogenide nanoparticles can be characteristics by the doping of the different metals which are like the holes. We present the effects of Pb and Sn doping on the electrical and thermoelectric properties of Tellurium Telluride $Tl_{10-x}Pb_xTe_6$ and $Tl_{10-x}Sn_xTe_6$ (x = 1.00, 1.25, 1.50, 1.75, 2.00) respectively, which were prepared by solid state reactions in an evacuated sealed silica tubes. Structurally, all these compounds were found to be phase pure as confirmed by the x-rays diffractometery (XRD) and energy dispersive X-ray spectroscopy (EDS) analysis. The thermo-power or Seebeck co-efficient (*S*) was measured for all these compounds which show that *S* increases with increasing temperature from 295 to 550 K. The Seebeck coefficient is positive for the whole temperature range, showing p-type semiconductor characteristics. Similarly the electrical conductivity (σ) and the power factors have also complex behavior with *Pb and Sn* concentrations. The power factor ($PF = S^2\sigma$) observed for $Tl_{10-x}Pb_xTe_6$ and $Tl_{10-x}Sn_xTe_6$ compounds are increases with increase in the whole temperature range (290 K–550 K) studied here. Telluride's are narrow band-gap semiconductors, with all elements in common oxidation states, according to $(Tl^{+})9(Pb^{3+})(Te^{2-})6$ and $(Tl^{+})9(Sn^{3+})(Te^{2-})6$. Phases range were investigated and determined with different concentration of *Pb and Sn* with consequents effects on electrical and thermal properties.

Keywords: Pb and Sn doping, Seebeck coefficient, electrical conductivity, power factor

1. Introduction

The thermo-electro-materials are now used as the renewable energy. It is used as the place of the coal, water tides, solar cells etc. The thermo electro-materials have more efficiency and reliable. Thermoelectric is one of the most important approaches in the solid state physics which can be converted the heat energy in the electrical energy, help to increase the efficiency, effectiveness and competency. It's importance is increase since last twenty years when the ease of use of fossil fuel is decrease. So there are different thermoelectric materials are used for the different temperatures from 10 K to the 1000 K which are used in the different applications for the cooling and heating [1–5]. Tellurium telluride is one important compound of the thermoelectric material which is studied, modified and increases the efficiency for the more and more applications for generation of power [1] and solar cells [2]. Tellurium telluride is a basically alloy that is used for the increases the energy conversion efficiency at the any temperature of the heating and cooling in the electrical circuit [3–7].

The figure of merit is

$$ZT = \frac{S^2 \sigma T}{k} \tag{1}$$

Where is σ is the electrical conductivity, k is the thermal conductivity, S is the see-beck coefficient, and T is the absolute temperature which is determined the efficiency of the thermo electric materials applications [8]. The power factors can be determined the electrical and thermal properties. The power factor can be defined as $S^2\sigma$. It can be help the determination of the charge carrier's concentration, from the doping concentration charges and lay down the free electrons in the system of chalcogenides.

We have investigated the chalcogenide with different materials (lead, tin, bismuth etc.) doped in the tellium tellurides. They have complex composition and structure on the basis of the electronic configurations. These compositions help to increases their properties like thermal, electrical, optical etc. of the thermoelectrical materials. There are many challenges of complex composition to high their electrical conductivity, high see-beck coefficient and low thermal conductivity. Due to this, they can controlled the electronic structures of the system i.e. band gap, shapes and degenerated level which is near the Fermi level, concentration of electrons and charge carriers scattered depend on them [7, 8].

Here is defined the basic chemical properties and are related by physical properties of different i.e. S, Se, and Te which are compared in **Figure 1**. Better rate performance and higher utilization rate of active materials can be realized in Te-based batteries. It is worth noting that Te delivers lower gravimetric capacity (419 mAh g^{-1}) than S (1672 mAh g^{-1}) and Se (675 mAh g^{-1}); this disadvantage can

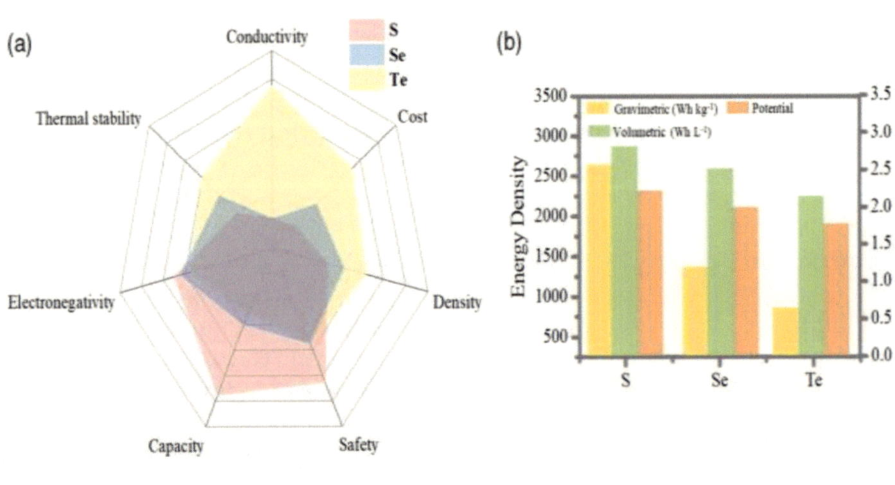

Figure 1.
(a) Comparison of S, Se, and Te based on various properties. (b) Energy density and plateau voltage of Li-S, Li-Se, and Li-Te batteries.

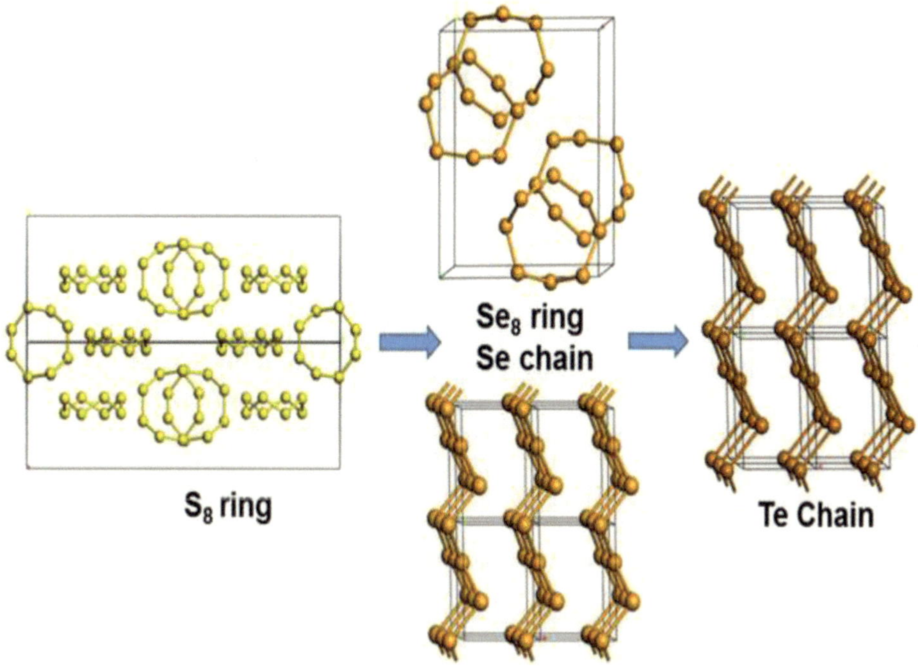

Figure 2.
Crystal structure transformations of S, Se, and Te. S crystal is constructed by S8 rings. Se crystals contain either Se_8 rings or Se chains. Te crystal only has chain-like structures. The difference in structure indicates that three congener elements exhibit distinctive structure changes during lithiation [16].

be overcome in volumetric capacity, which mainly results from its high density (2621 mAh cm^{-3} based on a density of 6.24 g cm^{-3}, compared with S 3416 mAh cm^{-3} and Se 3246 mAh cm^{-3}) [9–12]. Indeed, volumetric capacity is playing a more important role in practical application of batteries, resulting from the limited space in portable electronics [10, 13]. Considering the aforementioned, Te can serve as a high-performance electrode material in modern energy storage systems due to its high volumetric capacity and high electronic conductivity [14, 15].

It can be attributed to the different molecular arrangement of S, Se, and Te. S crystal is constructed by S8 rings. Se crystals contain either Se8 rings or Se chains. However, Te crystal only has chain-like structures, which shows infinite helix structure (**Figure 2**) [17]. There are huge difference makes a fundamental effect on poly-sulfides, poly-selenides, and poly-tellurides. At the second step of dis-charge process, poly-tellurides are continuously reduced to the insoluble metal di-tellurides and/or tellurides (M2Te2 or M2Te, M: Li, Na, or K) [18, 19]. Interestingly, this mechanism is only applicable to alkali metal-Te batteries; Al-Te batteries are working based on a different mechanism, which will be discussed later. The basic reaction mechanism of Te can be described as follows: during discharge, the Te positive electrode is first reduced to chain-like poly-tellurides and cyclo poly-tellurides (M2[Te*n*]2−, $2 < n \leq 8$, M: Li, Na, K) at the first step, which is different from metal-S and metal-Se batteries (only chain-like poly-sulfides or poly-selenides are generated) [11].

2. Experimental

The Sn and Pb doped $Tl_{10-x}A_xTe_6$(A = Sn &Pb) is (x = 1.00, 1.25, 1.50, 1.75, 2.00) has been prepared by solid state reactions in evacuated sealed silica tubes. The purpose of this study were mainly for discovering new type of ternary and quaternary compounds by using Tl^{+1}, $Sn^{+3}$and Te^{-2} elements as the starting materials. Direct synthesis of stoichiometric amount of high purity elements i.e. 99.99% of different compositions have been prepared for a preliminary investigation. Since most of these starting materials for solid state reactions are sensitive to oxygen and moistures, they were weighing stoichiometric reactants and transferring to the silica tubes in the glove box which is filled with Argon. Then, all constituents were sealed in a quartz tube. Before putting these samples in the resistance furnace for the heating, the silica tubes was put in vacuum line to evacuate the argon and then sealed it. This sealed power were heated upto 650 C^o at a rate not exceeding 1 k/mint and kept at that temperature for 24 hours. The sample was cooled down with extremely slow rate to avoid quenching, dislocations, and crystals deformation.

Structural analysis of all these samples was carried out by x-rays diffraction, using an Inel powder diffractometer with position-sensitive detector and CuKα radiation at room temperature. No additional peaks were detected in any of the sample discussed here. X-ray powder diffraction patterns confirm the single phase composition of the compounds.

The temperature dependence of Seebeck co-efficient was measured for all these compounds on a cold pressed pellet in rectangular shape, of approximately 5x1x1 mm^3 dimensions. The air sensitivity of these samples was checked (for one sample) by measuring the thermoelectric power and confirmed that these samples are not sensitive to air. This sample exposes to air more than a week, but no appreciable changes observed in the Seebeck values. The pellet for these measurements was annealed at 400 C^0 for 6 hours.

For the electrical transport measurements 4-probe resistivity technique was used and the pellets were cut into rectangular shape with approximate dimension of 5x1x1 mm^3. The proposed reaction equations are listed as follows.

3. Chemical reactions

First step:

$$\left(\frac{N}{8}\right)Te_8 + 2e^- + 2M^+ \leftrightarrow M_2Te_n\left(2 < N \le 8,,,M = Li,,,Na,,,K,,,Al\,Or\,Tl\right) \quad (2)$$

Second step:

$$M_2Te_n\,(2\mathrm{n}-2)e^- + (2n-2)M^+ = nM_2Te(2 < n \le 8) \quad (3)$$

Example.
Al-Te Battries:
Positive Electrode:

$$Te + 2e^+ \leftrightarrow Te^{2-} \quad (4)$$

$$Te^{2-} + AlCl_4^- \leftrightarrow AlTeCl + 3Cl^- \quad (5)$$

$$nTe + 2AlCl_4^- \leftrightarrow Te_n\left(AlCl_4\right)_2 + 2e^- \quad (\mathrm{n} = 2,4,6,8,10) \quad (6)$$

$$Te_n\left(AlCl_4\right)_2+\left(7n-2\right)AlCl_4^- \leftrightarrow nTeCl_3.AlCl_4+3nAl_2Cl_7+\left(4\text{n}-2\right)e^- \quad (7)$$

Negative Electrode:

$$4Al_2Cl_7^-+3e^- \leftrightarrow 7AlCl_4^-+Al \quad (8)$$

Overall Reaction:

$$5Al_2Cl_7^-+Te^{2-} \leftrightarrow 2Al+7AlCl_4^-+TeCl_3.AlCl_4^- \quad (9)$$

4. Result and discussions

4.1 Structural analysis

X-ray diffraction is used for the structural analysis of the materials. It helps to determine the crystal structure and particle size. Several of $Tl_{10-x}XTe_6$ which have doped Pb and Sn in it. The X is the different doped element. Where X is Pb and Sn. It has the different concentration of it. The X-ray diffraction of $Tl_{10-x}XTe_6$ with different concentration of doping of Pb and Sn as in **Figure 1**. Due to the different concentration, their peaks are different shown in **Figure 1**.

The **Figure 2** shows the EDX of the $Tl_{10-x}XTe_6$, have the different concentration of the doping of the Pb and Sn in it. The EDX shows the composition of the compounds. It shows the Pb and Sn are present in it.

4.2 Physical properties

To determine the different concentration of the doping of the Pb and Sn in the compound, there is changing in the charges carries. So the doping is effect on the temperature. Due to this temperature, it is variant in the Seebeck coefficient (S) as shown in **Figure 3**. The Seebeck coefficient can determined the temperature gradient for 1 K. It shows that the positive Seebeck effect from the 300 K to 500 K, for all p type semiconductors whose have the high charge carrier concentration. The Seebeck is positive due the concentration of doping elements is increase. So

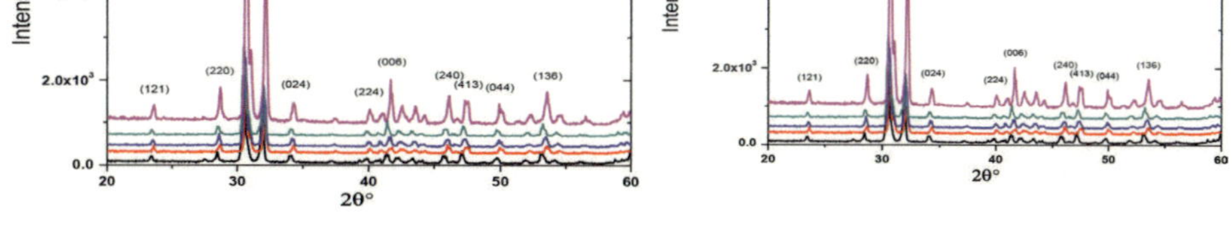

Figure 3.
XRD of doping of Sn and Pb in the TlXTe (X = Sn,Pb).

the mostly thermoelectric materials are the p type semiconductors materials. Due to increasing the concentration of doping elements, It improves the (i) reducing of grain size (ii) charge mobility and carrier density in thermos electric materials.

The Seebeck coefficient is varying from 80 to 120 μV/K as a function of the temperature. The behavior of the Seebeck coefficient is increasing as the Fermi level energy is decreasing due to the charge carrying density. In **Figure 4** shows that there is low level of charge carrier so that the holes are increase in it, so that it shows the high value of thermopower. So the large value of X, the doping elements have the large number of electrons and less number of charges carriers.

In **Figure 4** shows, the electrical conductivity of the quaternary compounds as compared to the temperature while the temperature is varied. The electrical conductivity is decrease as the temperature is increase that's why it is show the p type semiconductor and behave the positive temperature coefficient. It is cause the phonons scattering the charge carriers and effects the grains boundary. As increase the doping of elements, the holes in the compounds are increase, which is cause the phonons scattering. In chalcogenide system, the different elements are doping in the compound has no effect on the electrical conductivity. The low electrical conductivity is due to the effect oxide as the impurity in the compounds.

The behavior of temperature is different for the different concentration of the compound. The relationship between the Seebeck, temperature and concentration of doping elements as given below.

$$S = \frac{8\pi^2 k_B^2}{3eh^2} m^* T \left(\frac{\pi}{3n} \right)^{\frac{2}{3}} \qquad (10)$$

Where, k_B is the Boltzmann constant, e is the electronic charge, h is the Planck's constant, m^* is the effective mass and n is the charge carrier concentration. The effective mass and concentration are two parameters of the Seebeck coefficient. The samples have low concentration, it increase the thermos-power as well as the temperature.

The **Figure 4** shows that the electrical conductivity σ is decrease as the increase of the temperature of the compounds. The Seebeck is inversely effect due to the increasing of the doping of the concentration of the doping.

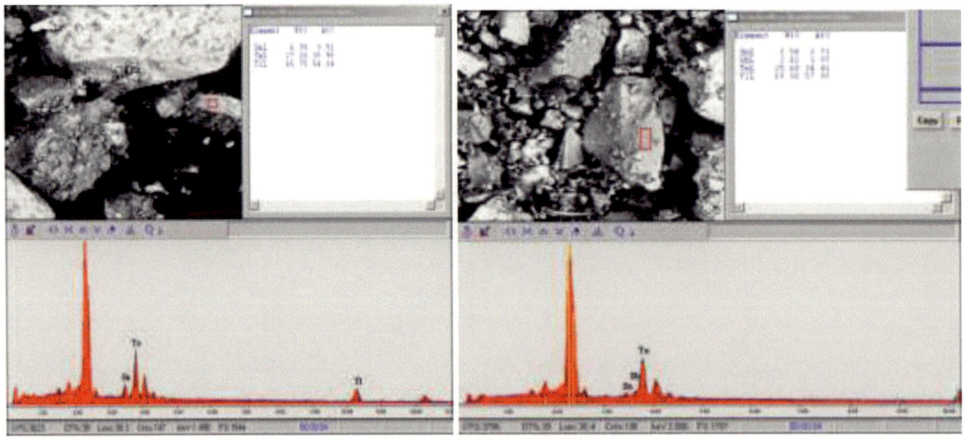

Figure 4.
Comparison of the EDX of doping of Pb and Sn in the TlXTe (X = Sn,Pb).

The different compounds have enhance the power factor (PF = $S^2 \sigma$) is decreases the electrical conductivity as increases the Seebeck coefficient in the given system. The PF is depend on the Seebeck coefficient. To measure the PF by the knowing the electrical conductivity and Seebeck coefficient in the **Figure 5**. As increases the temperature, the power factor is increases for all the compounds. In the **Figure 6** shows that the electrical conductivity is decreases as the increases the temperature is increases due the thermal vibration in compounds. The **Figure 7** shows that the power factor increase as the doping of the concentration is increases. As increases, the doping of the elements in the compounds is increase the optimization, which can help to increases the Seebeck and power factor (**Figures 5** and **7**).

Activity series of common metals

Li, K, Ba, Sr, Ca, Na	**Very active metals** **K and Na react violently with water** **They also react violently with acids.**
Mg, Al, Mn, Zn, Cr, Fe, Cd	**Metals of intermediate activity.** **React with steam or with acids such as HCl with liberation of H2**
Co, Ni, Sn, Pb, H2	**Moderately active metals.** **Do not react with water.** **React slowly with the HCl.**
Cu, Ag, Hg	**Moderately noble metals.** **Do not react with water, HCl.**
Pt, Au	**Very noble metals.** **React only with Aqua Regia.**

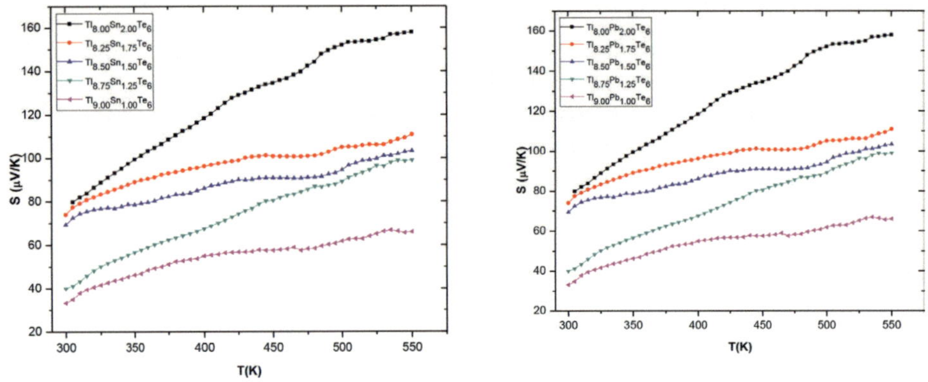

Figure 5.
Comparison of the see-beck coefficient of doping of Pb and Sn in the TlXTe (X = Sn,Pb).

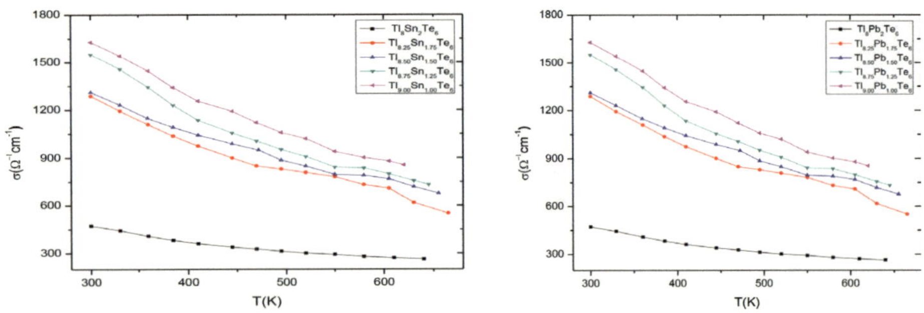

Figure 6.
Comparison of the electrical conductivity of doping of Pb and Sn in the TlXTe (X = Sn,Pb).

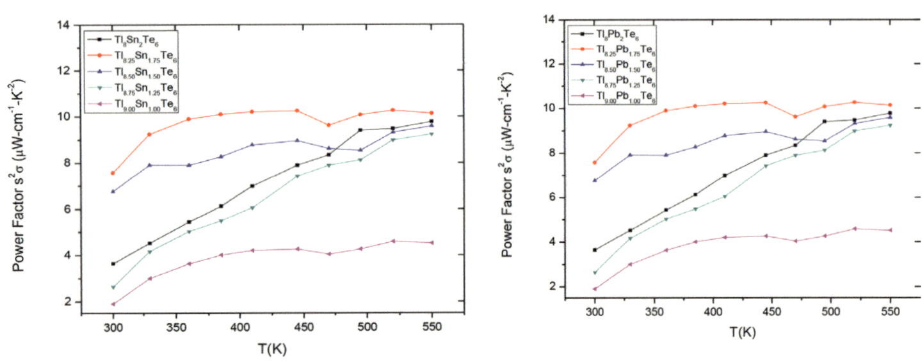

Figure 7.
Comparison of the power factor of doping of Pb and Sn in the TlXTe (X = Sn,Pb).

5. Conclusion

The different concentration of the doped Sn and Pb in the $Tl_{10-x}XTe_6$, is synthesized and then studied the physical properties. The XRD shows the compounds is the single phase, crystal structure measured by the experimental formula, having the same space group 14/mcm like Tl_5Te_3. The doping of the holes materials it changes its physical properties i.e. thermal, electrical, phase etc. Due the doping of Pb and Sn, the Seebeck coefficient is increases. The phase of the both compound

DOI: http://dx.doi.org/10.5772/intechopen.95390

is also change. The phase is come to the face centered cubic.it is also shows that the increases temperature decreases the electrical conductivity. The power factor is increases because the Seebeck coefficient is increase.

Author details

Waqas Muhammad Khan* and Wiqar Hussain Shah
Department of Physics, Faculty of Basic and Applied Science, International Islamic University, Islamabad, Pakistan

*Address all correspondence to: waqaskhanrwp@gmail.com

References

[1] T. Caillat, J. Fleurial, A. Borshchevsky, AIP conf. Proc. **420**, 1647 (1998).

[2] R.J. Campana, Adv. Ener. Conv. **2**, 303 (1962).

[3] R.J. Mehta, Y Zhang, C. Karthika eta, Nature Materials, **11**, 233-240 (2012).

[4] G.S. Nolas, J. Poon and M. Kanatzidis, MRS, Bull **31**199 (2006),

[5] B.A. Kuropaatawa, A. Assoud, H. Klienke, J. Alloys and Compounds **509**, 6768 (2011).

[6] K. Kurosaki, A. Kosuge, H. Muta, M. Uno, and S. Yamanaka, Applied Phys. Letts. **87**, 061919 (2005); J. Yang, F.R. Stablers, J. Electrn.Mater.**38**, 1245 (2009).

[7] G.J. Synder, E.S. Toberer, Nat. Mater. **7**, 105 (2008).

[8] J.R. Soostsman, D.Y. Chung, Kanatzdis, M.G. Angew Chem. Inter. Ed. **48**, 8616 (2009).

[9] R. Fang, S. Zhao, Z. Sun, D.-W. Wang, H.-M. Cheng, F. Li, Adv. Mater. 2017, **29**, 1606823.

[10] A. Eftekhari, *Sustain.* Energy Fuels 2017, **1**, 14.

[11] J. R. He, Y. F. Chen, W. Q. Lv, K. C. Wen, Z. G. Wang, W. L. Zhang, Y. R. Li, W. Qin, W. D. He, ACS Nano 2016, **10**, 8837.

[12] H. D. Jiao, D. H. Tian, S. J. Li, C. P. Fu, S. Q. Jiao, ACS Appl. Energy Mater. 2018, **1**, 4924.

[13] Y. Liu, J. W. Wang, Y. H. Xu, Y. J. Zhu, D. Bigio, C. S. Wang, J. Mater. Chem. A 2014, **2**, 12201.

[14] N. Ding, S.-F. Chen, D.-S. Geng, S.-W. Chien, T. An, T. S. A. Hor, Z.-L. Liu, S.-H. Yu, Y. Zong, Adv. Energy Mater. 2015, **5**, 1401999.

[15] S. Ullah, G. Yasin, A. Ahmad, L. Qin, Q. Yuan, A. U. Khan, U. A. Khan, A. U. Rahman, Y. Slimani, Inorg. Chem. Front. 2020, **7**, 1750.

[16] Y. Wang, H.-L. Fei, Ionics 2013, **19**, 771.

[17] J. R. He, W. Q. Lv, Y. F. Chen, K. C. Wen, C. Xu, W. L. Zhang, Y. R. Li, W. Qin, W. D. He, ACS Nano 2017, **11**, 8144.

[18] H. M. Li, K. L. Wang, H. Zhou, X. L. Guo, S. J. Cheng, K. Jiang, *Energy Storage* Mater. 2018, **14**, 267.

[19] S. Lee, H. Choi, K. Eom, J. Power Sources 2019, **430**, 112.

Section 2

Advanced Battery Technology

Chapter 4

Taming the Hydra: Funding the Lithium Ion Supply Chain in an Era of Unprecedented Volatility

Chris Berry

Abstract

The lithium ion supply chain is set to grow in both size and importance over the coming decade due to government-led efforts to decarbonize economies and declining costs of lithium ion batteries used in electronics and transportation. With forecasts of demand for lithium chemicals alone forecast to grow by three times later this decade, at least $10B USD is needed to flow into the upstream supply chain to ensure an efficient and timely build-out. Significant additional capital is needed for other portions of the supply chain such as other raw materials, cathode or anode production, and battery cell manufacturing. Recent exogenous shocks such as the US-China trade war and coronavirus disease 2019 (COVID-19) pandemic have made securing adequate capital for the supply chain a difficult challenge. Without the steady stream of funding for new mine and chemical conversion capacity, widespread adoption of electric vehicles (EVs) could be put at risk. This paper discusses the current structure of the lithium ion supply chain with a focus on raw material production and the need for and challenges associated with securing adequate capital in an industry that has, to date, not experienced such a robust growth profile.

Keywords: lithium, capital, capital allocation, investment, lithium ion, supply chain

1. Introduction

In Greek mythology, the hydra was a nine-headed monster that terrorized the local populace. When one of the hydra's heads was cut off, two grew back in its place. This evolving menace was ultimately killed by Heracles and his nephew Iolaus with Heracles severing each head and Iolaus cauterizing the wound, preventing the regrowth of additional heads.

In modern times, a problem described as a hydra is one which is multifaceted or continually evolving. The lithium ion supply chain has some similarities with the hydra in that there are multiple businesses (mining, refining, cathode and anode production, battery cell and pack production) which are all different in terms of operational complexity and funding needs. **Figure 1** demonstrates a high-level overview of the lithium supply chain.

Despite the complexity, the lithium ion supply chain is positioned for a strong growth based on the twin tailwinds of government requirements for decarbonization and falling lithium ion battery costs in the next decade. By the author's estimates, should electric vehicles (EVs) become 10% of global autosales later this decade, this would require three times more lithium than is currently produced

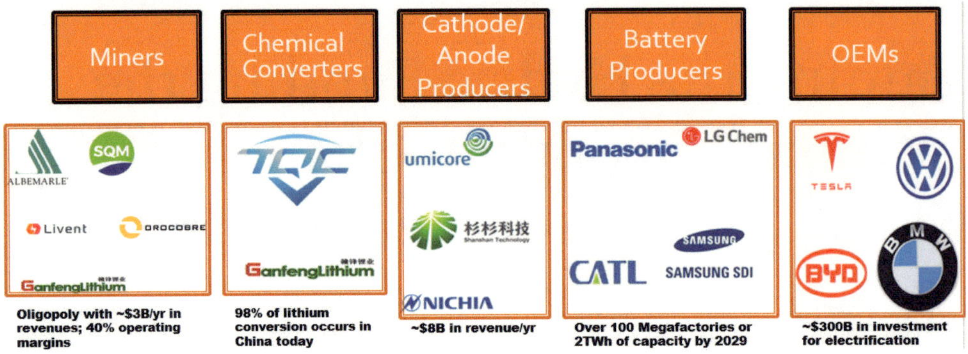

Figure 1.
A high level view of the lithium supply chain today. Source: House Mountain Partners, LLC.

globally today, given certain assumptions on battery size in kilowatt-hours (kWh) and battery chemistry. This tripling of demand ignores the growth in other sectors that use lithium such as pharmaceuticals or glass and ceramics.

However, the trajectory of this rather sunny scenario has recently been called into question. The US-China trade war and simultaneous supply and demand shock of coronavirus disease 2019 (COVID-19) are forcing investors and companies—traditional sources of capital to feed the growth of the supply chain—to pause and scrutinize capital allocation decisions. The entire industry is just beginning to understand the implications of these shocks, and this evolution can cause capital commitments to freeze or vanish altogether. In the past year alone, major lithium producers Albemarle, Sociedad Química y Minera de Chile (SQM), Livent, Ganfeng Lithium, and Tianqi Lithium (known collectively as "the Big Five" as they produce the majority of global lithium supply) have halted or slowed production expansion plans due to low lithium prices and softer than expected demand. The macroshocks referred to above have also hurt the development-stage mining companies with high-profile failures becoming more frequent.

With its multiple subsectors (mining, refining, cathode and anode production, and separator production, battery production, automotive), the comparison of the lithium ion supply chain to the hydra is apt.

The winds of change have dawned on one head of the hydra—the global automotive business. Mary Barra, *c*hief *e*xecutive *o*fficer (CEO) of General Motors (GM) referred to this volatility, stating in 2017:

> *I have no doubt that the automotive industry will change more in the next five to 10 years than it has in the last 50. The convergence of connectivity, vehicle electrification, and evolving customer needs demand new solutions [1].*

In an environment where equity investors view results on a quarterly basis, lithium company CEOs and chief financial officers (CFOs) are under immense pressure to deliver immediate returns to shareholders while at the same time ensuring that they maintain or increase market share by investing in the future and managing a lag in recouping those costs in the future.

This paper discusses the upstream lithium ion supply chain through the lens of capital allocation. The various subsectors of the business can be thought of as various heads of the ancient hydra—unwieldy, growing, and hungry (in this case for financial capital). I will discuss why governments around the world are intent on decarbonizing the transport sector, the importance of investment during different phases of the capital cycle, and the traditional challenges associated with capital allocation.

Trade war dynamics and COVID-19 have shocked the global economy and given investors pause in terms of how, when, and where to invest. The lithium ion supply chain is not immune here, but nobody disputes that in 10 years' time, this supply chain will be larger and more critical to the global economy than it is today. What is debatable is what the supply chain will look like and how to structure and deploy the enormous sums necessary for growth.

2. Why electrify?

The push for the electrification of transport via increased EV penetration centers on government goals to decarbonize the sector and mitigate the effects of climate change. At least 14 countries have announced their intention to ban the diesel engine by or before 2040 with numerous cities across the globe making similar announcements. The transport sector is responsible for 23% carbon dioxide (CO2) emissions and over 25% of oil demand according to the World Resources Institute (WRI) [2] and the US Energy Information Administration (EIA) [3]. The European Union (EU) has been a leader in the fight against automotive emissions, thanks in part to the Dieselgate scandal and concerns about climate change. The true global leader in pushing vehicle electrification to the fore, however, is China.

It is widely accepted among Chinese authorities that a key to continued growth of the domestic economy is through investment along multiple parts of next-generation supply chains, effectively owning the intellectual property that emerges from it. Programs such as Made in China 2025 have laid bare the desire on the part of the Chinese state for deeper integration into and dominance of next-generation supply chains. Made in China 2025 is essentially a blueprint for next-generation industrial dominance where China aims to control at least 70% of the global production of industries such as new energy vehicles, high-end microchips, renewable energy infrastructure, robots, and advanced medical technology with slightly less ambitious goals for other industries [4]. This push has met with predictable pushback from other competitor countries such as the United States, but China's intentions and goals are laid bare for all to see.

A second tailwind for electrification in the future concerns the lithium ion battery itself and its cost trajectory. Bloomberg New Energy Finance claims that battery prices in terms of $ per kWh have fallen by 87% since 2010 [5]. While Moore's law is not applicable with lithium ion batteries, these devices have undoubtedly benefitted from the technology-driven cost deflation which emanates from a mix of scaled production and technological advances in increased energy density. Lithium's light weight and capacity to store energy dictate that it is the material of choice for energy storage and mobility for at least the next decade despite other competing storage technologies. Lithium ion batteries have been commercially used for 30 years. That mix of a safety record, coupled with light weight and capacity to store a charge, places it at the top of the battery metal pyramid.

It is important to realize that metrics such as costs and demand growth rates will not move in a linear fashion as battery technology continues to improve. Though COVID-19 has caused a re-think of demand numbers across the supply chain, the general trend of double-digit demand growth remains intact over the long term. One of the challenges for automotive companies concerns the rapidly evolving battery technology and which will be the ultimate winner. There are at least six major types of lithium ion battery chemistries alone, and building a multibillion-dollar business around one or two of these chemistries is a decision many in corporate board rooms wrangle with as we speak. It is generally agreed upon in the industry that there will not be a single-cell format or lithium ion chemistry that ultimately wins, as different applications have different power and energy density requirements.

Regardless of the lithium ion chemistry of choice, the upstream portion of the lithium ion supply chain—raw materials—will need to grow rather dramatically in size in order to meet even base case assumptions for EV penetration in the coming years. This presents both an opportunity and a challenge as there are multiple lithium ion battery chemistries, multiple cell formats, and multiple battery sizes measured kilograms per kilowatt-hours. As an example, assuming a 10% EV sales penetration rate, the demand for lithium will increase by 3×, cobalt will increase by 2.5× based on author estimates. These numbers could change slightly based on changes in any of the factors previously listed, but the fact remains—we are going to need substantial amounts of additional raw material supply to electrify the global transport sector.

Similar demands will be placed on downstream portions of the supply chain with cathode production across different chemistries expected to grow strongly, cell and battery manufacturing growth all but obvious, and original equipment manufacturers (OEMs) committing reportedly $300B in electrification efforts across the next decade [6]. Each piece of the supply chain here must coordinate its growth with the other pieces (or heads of the hydra) in order to ensure optimal capital allocation and prevent waste and undue excess production capacity.

3. Supply chain growth and regionalization

For the lithium ion supply chain to grow sustainably as mentioned above, there is an assumption that needs to be made and several challenges that need to be navigated.

The assumption in question is that consumer demand for EVs and demand from fleets such as Amazon and car rental companies will materialize on a time frame suitable to investors and companies alike. This narrative still appears intact despite a COVID-19-led demand implosion and total cost of ownership (TCO) EV costs that are still higher than that of internal combustion engine (ICE)-powered cars. As of April 2020, it is still unclear what the effects of COVID-19 will be on consumer sentiment toward EVs, but some anecdotal evidence suggests that once price parity is achieved coupled with intangible benefits of cleaner air, the demand for EVs should recover [7]. This assumption bears careful scrutiny as the effects of COVID-19 continue to evolve. It would be fair to say that EV demand in the current environment will certainly be delayed, but not denied.

The benefits of companies such as FedEx or Amazon decarbonizing their fleets speak for itself as fleet operating efficiencies and the opportunities for companies to burnish their "green credentials" outweigh the near-term higher switching costs. Amazon has committed to having 100,000 electric delivery vans in its fleet by 2030, reducing carbon dioxide emissions by 4 million metric tons per year [8]. Again, widespread EV adoption rests on the assumption that these goals are achieved in the stated time frame.

The challenges to wider electrification have already been mentioned and perhaps deserve their own chapter in this book. While the lingering effects of COVID-19 are still to be determined, it is fair to conclude that smaller, more robust, regionalized supply chains are likely to result from current events. This effective "rebuild" of supply chains will take time, money, and political will. Combining all three is an enormous challenge, but there are examples to draw from historically such as the Marshall Plan which aided in rebuilding Europe in the wake of the Second World War. De-globalization is gaining appeal in 2020 as critical material dependence, income inequality, and the fragility of the existing global supply chain system are laid bare for all to see.

A new example of battery supply chain regionalization concerns the European Battery Alliance and its strategic action plan to position batteries at the center of a new wave of industrialization in Europe [9]. The underlying strategy of achieving a degree of autonomy and owning next-generation intellectual property is of paramount importance and a key measure of long-term success.

Future cash flows and battery-inspired intellectual property, also highly valuable, are at stake. The call for more self-sufficiency in supply chain development will only grow louder as globalization shrinks and the opportunity to dominate a fast-growing sector becomes apparent to investors of all stripes. It is to investors that we now turn.

4. Investment risks and implications

Perhaps the most overlooked aspect of lithium ion supply chain growth is the most important—the funding. Many people assume that the rollout to mass electrified transit will be linear and relatively seamless, but history suggests that transformative change is anything but. While discussions of double-digit compound annual growth rates (CAGR) for raw materials are enticing, pairing adequate sums of capital optimally along the supply chain has proven to be a challenging exercise. This is due to the fact that the mining segment of this supply chain is subject to infrequent and enormous cyclical swings in pricing, leading to a traditional boom-and-bust cycle. As an example of a typical commodity cycle is shown in **Figure 2**.

Embedded in the capital cycle thematic are risks unique to strategic metals such as lithium and cobalt. These can be broadly grouped into four categories: financial, geopolitical, social, and technology.

Financial risks center on issues like access to capital, determining the cost of capital, commodity pricing, the effect of technology on cost, and macroeconomic headwinds or tailwinds. The cost of and access to capital are causally related to what point we find ourselves in the capital cycle. When lithium carbonate pricing peaked at roughly $20,000 USD per ton in April 2018, equity share prices were at all-time highs, capital expansion plans were being announced on an almost daily basis, and capital was widely available. Fast forward to April 2020, equity share prices have crashed by as much as 65%, capital expansion plans have by and large

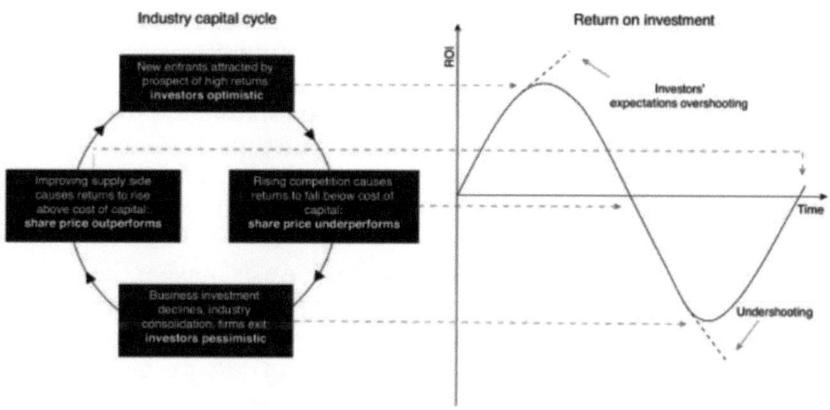

Figure 2.
The typical boom-and-bust cycle and return on investment. Source: Capital Returns: Investing through the Capital Cycle—A Money Manager's Reports 2002–2015, by Edward Chancellor.

been shelved by lithium miners and project developers, and the capital that is available has much more stringent strings attached to it than before to compensate for elevated risk. While some of this retrenchment is related to COVID-19, the asset behavior is typical of a boom-and-bust cycle.

Geopolitical risks include resource nationalism, resource dependence, and a single country (China, in this instance) exerting control over most of the full supply chain for critical minerals. When lithium pricing spiked through 2017, calls to protect "the people's lithium" from foreign exploitation in countries such as Chile only grew louder. The battery metals are somewhat unique in this regard as the small market size of lithium or cobalt (e.g., 300,000 tons per year and 140,000 tons per year, respectively) masks their strategic importance to next-generation technologies. Governments around the world such as the United States and the United Kingdom have designated these natural resources as "critical" and have proposed legislation to help lessen dependence on foreign adversaries [10] (**Figure 3**).

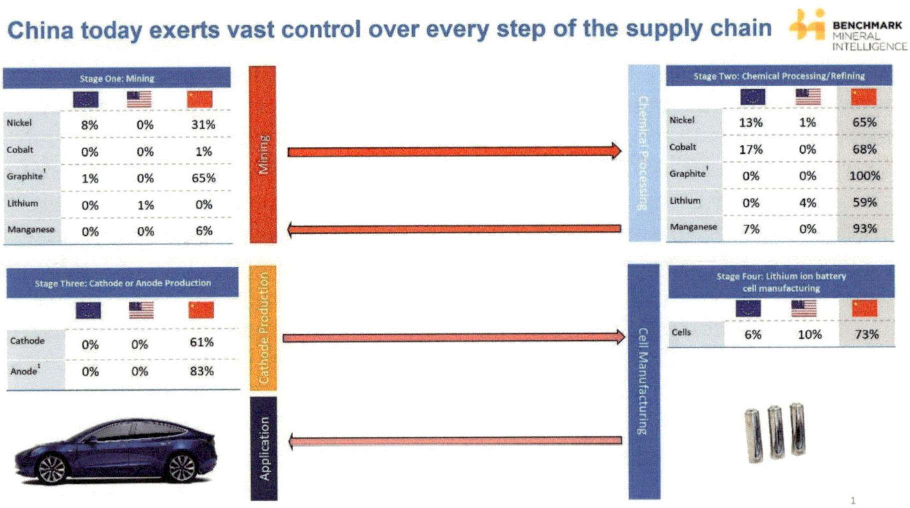

Figure 3.
China's dominance of all aspects of the EV supply chain is evident. Source: Benchmark Mineral Intelligence.

The social risks apply to the mining industry as a whole and have taken on new relevance with environment, social, and governance (ESG) mandates increasingly required by investors before committing any capital. What steps a company along the lithium ion supply chain is undertaking to minimize CO2 emissions is a notable example of renewed ESG focus. The recent questioning of globalization as an enabler of income inequality and the resulting explosion of populist sentiment is another example of the challenges lithium ion supply chain companies face amidst a backdrop of strong growth in the coming decade. How this growth will be financed given ESG mandates and how the wealth will be shared (if at all) are looming questions.

A final risk is centered on the role of technology. An example is the optimization and deployment of new technologies to lower costs in lithium extraction. Better known as direct lithium extraction (DLE) technologies, these technologies such as solvent extraction or ion exchange have the potential to reshape the lithium cost curve. While the technologies themselves are well known, the potential benefits to lithium include dramatically lower extraction and purification costs, mitigation of the need for large brine ponds, and reduction in the time it takes to produce high-purity lithium chemicals. While this technology is not scaled as of this writing, there are a number of these processes competing for investor capital predicated on these benefits [11].

4.1 The trouble with pricing

The recent price swings in lithium and cobalt have caused more damage to capital availability in the lithium ion supply chain than benefit. Raw material price volatility presents one of the greatest challenges to lithium ion supply chain development in the next decade. Enormous sums are required to build additional production capacity, but excessive volatility in raw material pricing has typically kept all but the largest strategic investors out of the space. These strategic investors—mainly Asia-based banks or lithium and cobalt users—only have so much capital and so much capacity to process raw materials. A move to vertically integrate supply chains is believed to be the answer from the automakers, but this process will be somewhat slower than many would like as automotive supply chains as they exist today will not be able to "turn on a dime" and ramp production of EVs owing to the sunk costs of existing infrastructure. Perhaps the vertical integration strategy of Tesla or China's view of owning most of the supply chain serves as a viable model going forward.

For the sake of reference, the typical greenfield lithium mine can have an upfront capital expenditure of $500–600 M USD (inclusive of the mine, ore type, and processing plant) [12]. From discovery of the asset, through development, successful capital raising, deployment of the said capital, project build-out, and first production, the typical timing is 7–10 years. If consensus lithium demand is anticipated to grow by three times in the next decade, this would require at least $10B USD in capital in an industry that traditionally generates only $3B USD in revenue per year. Should EV demand accelerate ahead of consensus expectations, this $10B figure could be conservative.

A challenge here for capital providers is that in addition to effectively timing the market and managing anticipated demand, the pace of change is accelerated when confronted with disruptive technologies such as battery chemistries or lithium extraction technologies. While lithium demand can increase at a linear rate, technological leaps can accelerate this demand just as COVID-19 can destroy it and wreak havoc with pricing.

Lithium ion battery chemistries such as lithium iron phosphate (LFP) are expected to lose market share quickly to their more energy dense cousins nickel-manganese-cobalt (NMC) and nickel-cobalt-aluminum (NCA). This has not come to pass yet however for several reasons including improvements in LFP energy density [13] and lower overall raw material pricing which affects switching costs.

Aside from questions surrounding battery chemistries, the crash in equity share prices in Q1 2020 has served to push the typical investor into a "risk off" mentality despite global central banks effectively trying to backstop the financial markets and prop up asset prices using both monetary and fiscal levers. Without a healthy risk appetite in the markets, how will the CO2 reduction targets in the EU be achieved in time? How will China achieve its goal of 20% of new car sales in 2025 as EVs? The reality is that for these goals to be achieved, a flood of capital needs to come into the lithium ion supply chain in a hurry. Otherwise, the EV market will remain smaller for longer, with fewer winners who were initially willing to embrace risk and invest in supply chain development.

4.2 Visibility through futures contracts

The lithium industry has come together recently to produce a solution for the lack of ability to hedge price risk. The London Metal Exchange convened a working group in 2019 to build futures contracts for lithium chemicals [14]. These contracts are aimed at providing price transparency across what has traditionally been a small, opaque market run in an oligopolistic fashion. The contracts are to be cash settled and will source pricing data from the members of the working group. As of April 2020, the contracts had yet to be introduced owing to logistical issues with price collection. Pricing data is highly proprietary in nature, and the futures contracts need this critical component to ensure widespread adoption across the industry. The opacity of pricing works to the advantage of lithium chemical producers such as Albemarle or SQM but acts as a detriment to potential capital providers who need pricing transparency and stability to commit capital at a reasonable cost for a significant length of time.

Owing to the traditionally small size of the market and highly specialized end products, lithium pricing has always been opaque with pricing determined on a contracted basis between the producer and buyer with a small spot market which exists in China for "off spec" material. Contract pricing between the producer and seller is a major piece of off-take agreements and is really driven by four criteria: the term of the contract, the amount of material for purchase, the quality of the material, and the depth of the relationship between producer and consumer. There can be a wide degree of flexibility in terms of typical off-take agreements in order to protect both producer and consumer from price volatility or other events which could require force majeure. A futures contract can help serve as an anchor to any off-take agreements.

5. Shared risks in capital allocation: a case study

While the immediate need for large sums of capital is clear, the source of this capital has traditionally been centrally located with Asia-based banks or end users such as battery manufacturers serving as off-take partners. While Asia, and China in particular, will remain the main engine of demand for EVs, other parts of the world such as the European Union are likely to become a larger part of the EV sales "pie." This would indicate that sources of capital will need to become more diverse and perhaps serve local markets first. Given the typically large sums needed for initial capital expenditures, much of the capital comes in the form of debt instruments along with an equity component to ensure all stakeholder interests are aligned. The idea of trading offtake for financing has become a trusted model in the industry as the end users of lithium chemicals, for instance, typically have the robust balance sheets necessary to help with the financing of mining projects. One such case study is Pilbara Minerals, a lithium producer based in Australia. The company has executed several off-take agreements, but a good case study is the lithium concentrate agreement the company completed with Ganfeng lithium in 2017 [15]. This is far from the only off-take agreement Pilbara has executed, but what makes this deal unique is that it has multiple components. Ganfeng agreed to purchase up to 160,000 tpy of lithium concentrate, as well as taking a $20 M AUD equity position in Pilbara and offered a debt funding/pre-payment facility to alleviate the financing challenges. This partnership model, where risk is spread across all stakeholders and both financing and technology challenges are mitigated, has been used with other lithium producers and off-take partners with success.

5.1 Capital allocation gone wrong: a case study

Lithium is an oligopoly, with Albemarle, SQM, Livent, Ganfeng Lithium, and Tianqi Lithium as the major lithium compound producers. During the most recent bull market in lithium (2016–2018), these companies all pushed forward with project expansions, and M&A activity was scarce with one exception. Tianqi Lithium purchased a 24% equity stake in SQM in 2018 for $4.066 billion USD. This deal raised several eyebrows in the lithium world in the sense that two fierce competitors were integrating their respective strategies with SQM giving up more in terms of market intelligence, pricing, and sales strategies and gaining much less. Tianqi's strategy was clear—the share purchase instantly made the company a major SQM shareholder and three board of director seats included with the equity position would ensure an intimate view of SQM's marketing strategy and expansion plans—plans that Tianqi could exploit to their own advantage. The deal was enveloped in controversy from the start in Santiago and ultimately while the transaction was completed successfully, a number of restrictions were placed on the three board seats [16].

The real capital allocation lesson from this deal, however, had yet to be learned. When Tianqi purchased the SQM Class A shares, they did so at a price of $65 per share. The purchase was funded in part with a $3.5B USD loan from CITIC Bank in China. Since the deal was completed, SQM share price has fallen by 65%. This, coupled with a dramatically lower lithium chemicals price, has caused severe liquidity issues for Tianqi, harming their credit rating which was subsequently downgraded to Caa1 with a negative outlook by Moody's in March of 2020 [17]. Adding a sense of urgency, around $2B of the loan is due in November 2020. Essentially, Tianqi made a deal at the absolute top of the capital allocation cycle and must now work to strengthen its balance sheet at a particularly precarious and unpleasant point in lithium industry history. Unfortunate allocation decisions such as this harm investor confidence in the lithium sector and could delay much-needed funding for expansion.

6. Conclusion

The lithium ion supply chain has experienced twin shocks of a trade war and COVID-19 demand destruction against a backdrop of strong long-term growth dynamics. These forces have kick-started a re-think of globalized supply chains in the interest of national security and self-sufficiency. Even those who are bearish on EV demand see the market growing throughout this decade, and so it makes sense to ensure that supply chains are built with sustainability and resilience in mind as raw material demand experiences robust growth. Recent actions by the US, EU, and Chinese governments are but a taste of what is to come.

The decarbonization thematic goes together with de-globalization and is benefitting from the twin tailwinds of global government mandates to phase out the ICE and tech-driven cost deflation in the cost of lithium ion batteries. These forces will not remain in place indefinitely but however are set to create opportunities for a wider swath of investors. Assuming the rosy demand projections come to fruition, a robust and resilient lithium ion supply chain depends upon two things: successful capital accumulation and capital allocation. A world awash in liquidity indicates that capital accumulation in really a function of expected returns on given project and the best projects will attract enough capital to sustain themselves through the building and commissioning phases. The challenge comes with capital allocation, and there are many lessons learned here about how not to deploy capital rather than the opposite. In recent years, capital in the lithium ion supply chain has flowed from a small coterie of strategic investors given raw material price opacity, price volatility, and the generally small size of the markets. This will have to change if we are to meet most EV penetration forecasts in the future.

The lithium supply chain faces a dilemma. If decarbonization and EV sales penetration goals are to be met, much more capital is needed to fund mine and supply chain development today for a demand profile that only exists on paper. The checkered history of lithium mine capital accumulation and allocation does not lend confidence to those looking to invest, so how deals are structured will be a key to unlocking new sources of capital.

The unwieldy and unprecedented hydra-like growth of the lithium ion supply chain will offer multiple exciting opportunities to alter supply chains and achieve sustainable results for all stakeholders well into the future. The lynchpin for success is successful capital accumulation and deployment.

Acknowledgements

The author wishes to thank Dr. Michael A. Berry, Simon Moores of Benchmark Mineral Intelligence, and Alex Grant of Jade Cove Partners for their assistance in the writing of this chapter.

Conflicts of interest

The author declares no conflicts of interest.

Author details

Chris Berry
House Mountain Partners, LLC, New York, USA

*Address all correspondence to: cberry@house-mountain.com

References

[1] General Motors Corporate Communications. 6 January 2016. Available from: https://media.gm.com/media/us/en/gm/autoshows/detroit.detail.html/content/Pages/news/us/en/2016/Jan/boltev/0106-barra-ces.html

[2] Wang S, Ge M. 16 October 2019. Available from: https://www.wri.org/blog/2019/10/everything-you-need-know-about-fastest-growing-source-global-emissions-transport

[3] U.S. Energy Information Administration. 2019. Available from: https://www.eia.gov/outlooks/ieo/pdf/transportation.pdf

[4] Institute for Security and Development Policy. June 2018. Available from: https://isdp.eu/content/uploads/2018/06/Made-in-China-Backgrounder.pdf

[5] Bloomberg New Energy Finance. 3 December 2019. Available from: https://about.bnef.com/blog/battery-pack-prices-fall-as-market-ramps-up-with-market-average-at-156-kwh-in-2019/

[6] Leinert P, Shirouzu N, Taylor E. 10 January 2019. Available from: https://www.reuters.com/article/us-autoshow-detroit-electric-exclusive/exclusive-vw-china-spearhead-300-billion-global-drive-to-electrify-cars-idUSKCN1P40G6

[7] Berman B. 9 April 2020. Available from: https://electrek.co/2020/04/09/study-pandemic-lockdowns-are-encouraging-more-consumers-to-buy-electric-cars/

[8] Shieber J. 19 September 2019. Available from: https://techcrunch.com/2019/09/19/amazons-climate-pledge-commits-to-net-zero-carbon-emissions-by-2040-and-100-renewables-by-2030/

[9] Commission to the European Parliament, The Council, The European Economic and Social Committee, The Committee of the Regions, and The European Investment Bank. April 2019. Available from: https://ec.europa.eu/commission/sites/beta-political/files/report-building-strategic-battery-value-chain-april2019_en.pdf

[10] The White House. 20 December 2017. Available from: https://www.whitehouse.gov/presidential-actions/presidential-executive-order-federal-strategy-ensure-secure-reliable-supplies-critical-minerals/

[11] Grant A. Disruption or Evolution. Benchmark Mineral Intelligence Q1, 2020 Review. 2020

[12] Piedmont Lithium Annual Report. 2019. Available from: https://d1io3yog0oux5.cloudfront.net/_b3420a839348465b7684789d134ca977/piedmontlithium/db/338/2563/pdf/Piedmont+Australian+Annual+Report_2019_Merged_FINAL.pdf p. 4

[13] BYD Press Release. 29 March 2020. Available from: https://en.byd.com/news-posts/byds-new-blade-battery-set-to-redefine-ev-safety-standards/

[14] Available from: https://www.lme.com/About/Corporate-information/Committees/Lithium-committee [Accessed: April 2020]

[15] Pilbara Minerals Press Release. 2 May 2017. Available from: http://www.pilbaraminerals.com.au/site/PDF/7116c544-760c-4758-a395-72a987e59c31/OfftakeandFinancingSupportwithGanfengLithium

[16] SQM Press Release. 17 May 2018. Available from: http://s1.q4cdn.com/793210788/files/doc_news/2018/5/PR_Nutrien_17May18.pdf

[17] Moody's Investor Service Press Release. 25 March 2020. Available from: https://www.moodys.com/research/Moodys-downgrades-Tianqi-Lithium-to-Caa1-outlook-negative--PR_419977?WT.mc_id=AM%7ERmluYW56ZW4ubmV0X1JTQl9SYXRpbmdzX05ld3NfTm9fVHJhbnNsYXRpb25z%7E20200325_PR_419977

Chapter 5

Ionic Liquid-Based Gel Polymer Electrolytes for Application in Rechargeable Lithium Batteries

Himani Gupta and Rajendra K. Singh

Abstract

Depleting fossil fuels has put pressing need for the search of alternative energy resources. Solar and wind energy resources are being considered one of the viable solutions. However, these intermittent sources require efficient energy storage systems in terms of rechargeable Li batteries. In Li batteries, electrolyte is one of the most important components to determine the performance, as it conducts the ions between the electrodes. In battery, mostly liquid electrolyte is used as it shows high ionic conductivity and electrode/electrolyte contact which help to reduce the internal resistance. But these are not electrochemically very stable and raised some major problems such as reactivity with electrode, dissolution of electrode ions, leakage, volatility, fast Li dendrite growth, etc. Therefore, in order to improve its electrochemical performance, selection of electrolyte is an important issue. In the present study, ionic liquid (IL)-based polymer electrolyte is used over liquid electrolyte in which IL acts as a plasticizer and improves ionic conductivity and amorphicity. These electrolytes have high thermal and electrochemical stability, therefore, can be used in high voltage Li battery. Also, their mechanical stability helps to suppress Li dendrites growth. Therefore, polymer electrolytes can open a new way in the progression of battery application.

Keywords: ionic liquid, polymer electrolyte, ionic conductivity, electrochemical analysis, Li battery

1. Introduction

Energy contributes a major role in various aspects of human life and their demand is continuously increasing with time. To meet this challenge, world population mostly depends on fossil fuels which include petroleum, coal, nuclear power and natural gases. But the climate change due to the frequently increasing CO_2 level has driven the research in the development of renewable energy sources. Research in the field of renewable energy sources such as solar energy, wind energy and hydropower has been initiated but the major problem to efficiently utilize these energies is to investigate suitable electrical energy storage devices because these energy sources are intermittent in nature. For this purpose, the most efficient energy storage devices are the batteries and super capacitors [1, 2]. Both the devices have their own importance depending on usage but batteries are continuing to dominate in the market of portable electronic system because of their high energy density and voltage rating [3]. Battery is an electrochemical device which can convert chemical energy into electrical energy through redox reaction to release the energy. Conversely, it can also convert electrical energy into chemical energy to store the energy. Among all the batteries, rechargeable lithium batteries (Li-batteries) are gaining much attention in the electric power storage system due to their high capacity, working voltage, long lifetime, low self-discharge rate and no memory effect (**Figure 1**) [4, 5]. However, the energy density of recent Li batteries is in the range of 100–200 Wh/Kg which limits their use in automotive application [6]. Therefore, to enhance their energy density, Li metal is frequently used because it shows very high capacity for Li battery. Metallic lithium as anode is being used since long time but its application with organic liquid electrolytes arises the main issue in lithium batteries as lithium dendrite growth [7–9]. Further, the use of these volatile and flammable electrolytes causes safety problems during cycling. Also, these organic liquid electrolytes cannot be used in high voltage batteries due to their electrochemical instability at higher voltage [10].

Therefore, for safely utilization of Li metal in batteries, requirement of alternative electrolyte is highly demanded. In this context, polymer electrolytes are gaining much attention in Li batteries because of their outstanding properties such as mechanical, thermal and electrochemical stability, safety and flexibility [11, 12]. The polymer acts as a host matrix for ion movement in which ions can move in the free space provided by the polymer matrix. Generally, solid polymer electrolytes (SPEs) are formed by dissolving organic salt into the polymer matrix. The selection of polymer matrix mainly depends on the presence of polar group so that it can easily coordinate with the cations; and there is less restraint in the bond rotation [13]. Among the different polymer matrices, poly (ethylene oxide) (PEO) based polymer electrolytes are mostly studied due to its high chain flexibility and ability to dissolve different organic/inorganic materials [14, 15]. However, PEO based solid electrolytes are semi-crystalline in nature which comprise the crystalline and

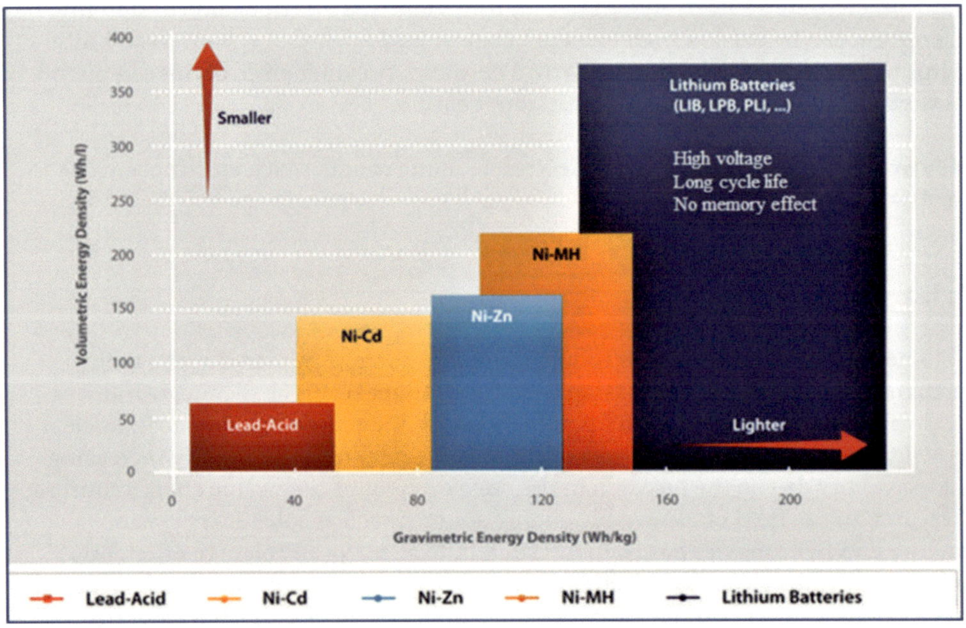

Figure 1.
Comparison in different types of batteries.

amorphous regions both; and it is reported that the ion conduction in polymer electrolytes occurs only in amorphous region [16–18]. Therefore, they show lower room temperature ionic conductivity (10^{-8} to 10^{-6} S/cm) and hence cannot be used in practical battery application. To overcome this drawback and to obtain polymer electrolytes having good ionic conductivity, flexibility, mechanical and electrochemical stability, different approaches are reported such as the addition of nanofillers, use of plasticizers or gel polymer electrolytes (GPEs) [19, 20]. Among these approaches, use of GPEs is in focus as they combine the desirable properties of both liquid (high conductivity) as well as solid (mechanical stability). Thus it is a suitable replacement of electrolyte for high performance batteries. In the present chapter, fundamental properties of PEO based polymer electrolytes; their classifications and performance in Li batteries are discussed.

2. Properties and ion transport mechanism in PEO

PEO has chemical structure $H{-}(O{-}CH_2{-}CH_2)_n{-}OH$ which is a polyether compound. Depending on the molecular weight of polymer, it is also known as polyethylene glycol (PEG). Generally, polymer having molecular weight above 20,000 g/mol is called PEO while, below 20,000 g/mol is known as PEG [21]. PEO is a low toxic compound; therefore, it is widely used in many applications such as chemical, industrial, medical, biological etc. PEO has ethylene oxide unit which provides flexibility and high donor number for Li^+ ions and promotes the ion transport. Also, it has high dielectric constant and Li^+ ion solvating ability [22]. The ion transport mechanism in PEO occurs by the coordination of Li^+ ions with the ether oxygen atom of PEO chain. This process occurs by formation and dissociation of Li—O bond by local segmental motion of polymer chain (**Figure 2**) [23]. The ion transportation in polymer

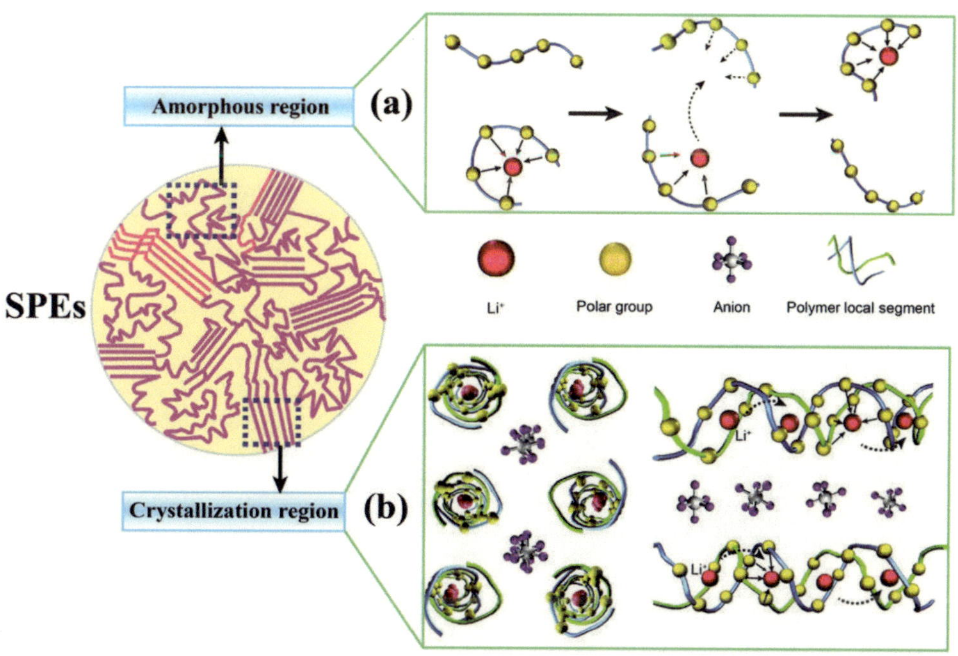

Figure 2.
Ion transport mechanism in polymer PEO in (a) amorphous as well as (b) crystalline region.

electrolytes mainly depends on polymer chain segmental motion which becomes faster in amorphous region or above the glass transition temperature (T_g) of polymer (as shown in **Figure 2(a)**). Hence, ideally polymer should have lower T_g to remain in rubbery state or to promote the Li^+ ion conductivity at RT [17]. In order to enhance Li^+ ion conductivity of polymer electrolytes different strategies such as addition of nano fillers, use of plasticizers or gel polymer electrolytes has been proposed. On this basis, polymer electrolytes are classified into different categories as discussed below.

3. Classification of polymer electrolytes

Polymer electrolytes are considered to be promising materials in the research and development of electrochemical devices. On the basis of materials, polymer electrolyte is classified into following categories (**Figure 3**):

3.1 Dry solid polymer electrolytes

It is formed by incorporating inorganic salt into the polar polymer thus ion conducting electrolyte is known as solid polymer electrolyte [24]. The electrostatic interaction between the metal ions of salt and polar polymer results the formation of coordination bond. This metal-polymer interaction can be affected by many factors such as nature and distance between the functional group of polymer, molecular weight, nature of branching, charge on metal and counter ion [25]. When the polymer electrolyte is placed in the electric field, ions start to move from one coordination site to other. It occurs due to weaker interaction between the metal ion and functional group of polymer chain.

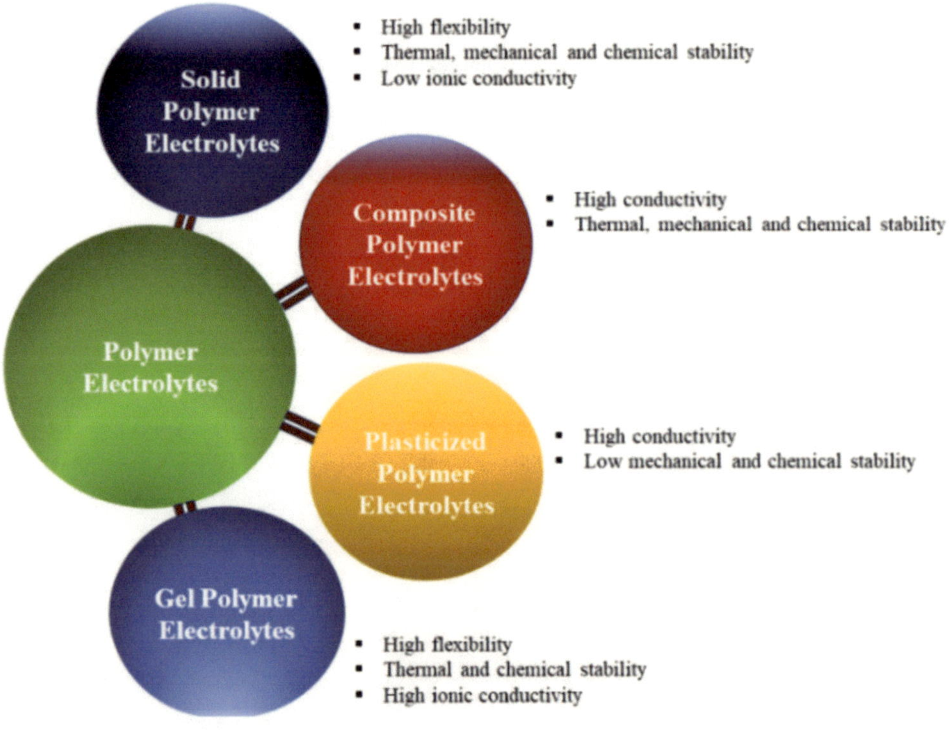

Figure 3.
Different types of polymer electrolytes used in Li battery.

3.2 Plasticized polymer electrolytes

Plasticized polymer electrolytes are formed by dissolving low molecular weight compounds (e.g. ethylene carbonate (EC), propylene carbonate (PC), poly ethylene glycol (PEG)) [26]. These plasticizers reduce the inter and intra-molecular interaction between the polymer chain, thus reduce the T_g and crystallinity of polymer chain and enhance the salt dissociation ability [27]. Although this approach improves the conductivity of polymer electrolyte, but it also provides low mechanical stability, solvent volatility and reactivity with lithium electrode.

3.3 Composite polymer electrolytes

Conductivity of polymer electrolytes also decreases due to the presence of ion pair formation. This behavior is observed mainly because of the low dielectric constant of polymer matrix [28]. To overcome this issue, high dielectric constant inorganic inert fillers (such as TiO_2, SiO_2, etc.) are dispersed into the polymer electrolyte to avoid ion-ion association. The obtained electrolyte combines the properties of ceramic fillers as well as polymer and results the flexible, mechanically stable composite polymer electrolyte [29].

3.4 Gel polymer electrolytes

In all the polymer based electrolytes, gel polymer electrolytes (GPEs) are focusing much attention because they combine the advantages of liquid electrolyte such as high conductivity, good electrode/electrolyte contact and solid electrolytes like as safety, mechanical and thermal stability. In GPEs, polymer is used to trap the liquid constituent and provides mechanical support, thus it is considered as safer than liquid electrolytes. For the synthesis of GPEs, large amount of organic solvent is used in the polymer in which its conduction takes place along the host polymer [30]. Recently, ionic liquid (IL) based GPEs are in focus of the research due to the desirable properties of ILs such as high conductivity, thermal stability, negligible vapor pressure which makes GPE as a suitable replacement of liquid electrolyte.

4. Ionic liquid

ILs are the molten salts that remain in liquid state below 100°C. Sometimes these are also referred as room temperature molten salts, ionic fluids, fused salts or organic salts. ILs are generally formed by self-dissociated, poorly coordinated, bulky organic cations and organic/inorganic anions [31]. Some of the common cations and anions of ILs are given in **Figure 4**. These ILs do not have strong ionic bond between the cations and anions as in ionic salts (NaCl, KCl, etc.), hence possess low lattice energy and remain in dissociated state. Therefore, they show many desirable properties as high conductivity, low vapor pressure, melting and glass transition temperature, high thermal and electrochemical stability, less polluting and easily recyclable [32]. Some of the properties of ILs are given in **Table 1** [33–35].

4.1 IL based gel polymer electrolytes

ILs provide outstanding ionic conductivity upto the decomposition temperature which enable them to be used as electrolyte. Also, to eliminate their leakage issue, ILs are trapped in polymer matrices which are known as GPEs. These IL based GPEs show high conductivity, good thermal and electrochemical stability, transparency and flexibility (**Figure 5**) [36].

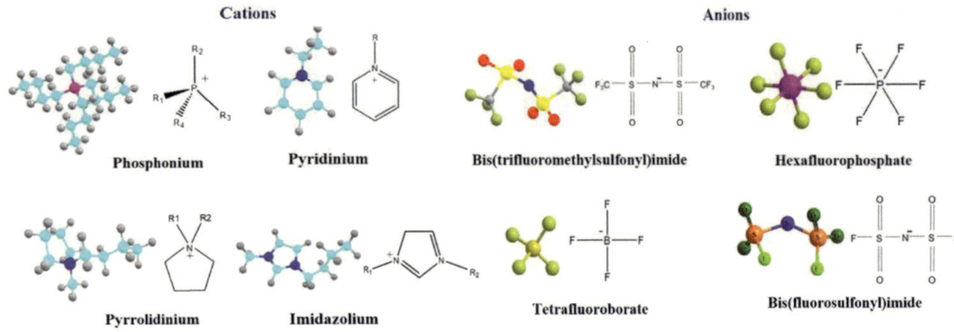

Figure 4.
Common cations and anions of ionic liquids.

Ionic liquids	T_g (C)	T_m (C)	T_d (C)	σ (mS/cm)	η (cP)
$PYP_{1,4}$-TFSI	-87	-6	—	2.69	60
[EMIM] [BF_4]	-93	-11	450	14 at 25°C	43 at 20°C
[EMIM] [PF_6]	—	60	—	5.2 at 26°C	—
[EMIM] [TFSI]	-98	4	440	8.8 at 20°C	28 at 25°C
[BMIM] [Cl]	—	41	254	—	1534 at 50°C
[EMIM] [FSI]	—	12.9	—	16.5 at 25°C	24.5 at 25°C
[HMIM] [PF_6]	-78	-61	417		585 at 25°C
[BMIM] [PF_6]	-76	10	390	1.8 at 25°C	312 at 25°C
[BMIM] [TFSI]	-104	-4	439	3.9 at 20°C	52 at 25°C
[OMIM] [PF_6]	-82	-40	376	—	682 at 25°C
[HMIM] [BF_4]	-82.4	-18	409	1.22 at 25°C	439

Table 1.
Properties of ionic liquid.

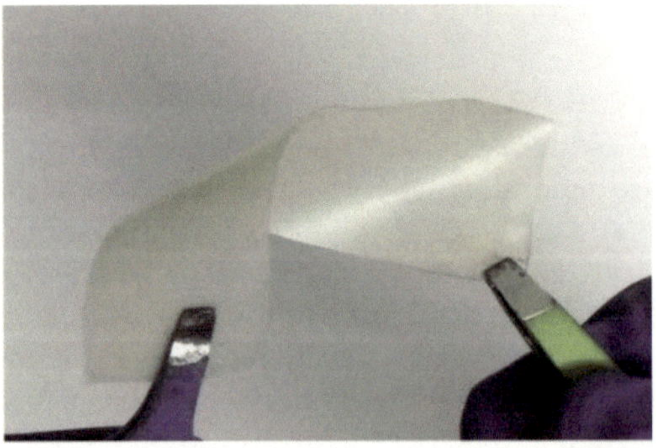

Figure 5.
Flexibility and mechanical stability of IL based gel polymer electrolyte.

They also provide good adhesive nature with the electrode surfaces. The transportation of ions in GPEs occurs by hopping or diffusion process. IL enhances the performance of polymer electrolyte in two ways (i) it acts as the plasticizer therefore reduces the crystalline phase or enhances the amorphous region, (ii) Supplier of free charge carriers, hence helps to improve the ionic conductivity of polymer electrolytes [19]. The ionic conductivity of different IL based GPEs are shown in **Table 2**.

ILs based GPEs	Conductivity (mS/cm)	References
$PEO_{20}LiTFSI_2[Pyr_{1.201}TFSI]_4$	2.5×10^{-4} at 20°C	[37]
$PEO_{20}LiTFSI_1[Pyr_{1.201}TFSI]_{1.5}$	7×10^{-5} at 20°C	[35]
PVdF-HFP + 20 wt.% LiTFSI+60% $BMIMBF_4$	1.7 at RT	[38]
$PPEGDA_{15\%}[LiBF_4Im_{12}BF_4]_{85\%}$	1.2×10^{-4} at 20°C	[39]
$[PPyr11TFSI]_{50\%}[Li(G_4)]TFSI_{50\%}$	1×10^{-4} at 20°C	[40]
EGDMA-MMA + 0.5 M $PP_{14}Cl$ + 80% $PP_{14}TFSI$	0.09 at 25°C	[41]
PEO + 20wt%LiFSI+7.5wt%EMIMFSI	0.289 at RT	[36]
PEO + 20wt%LiTFSI+12.5wt%EMIMTFSI	0.208 at 30°C	[42]
$PEO_{20}LiTFSI\ [Pyr_{13}TFSI]_{2.15}$	3×10^{-4} at 20°C	[43]
PEO + 10wt%NaMS+60wt%BMIM-MS	0.105 at 30°C	[44]
PEO + 10% NaTFSI+40% BMIMTFSI	0.4 at 30°C	[45]
$PEO_{20}LiTFSI_2\ [Pyr_{14}TFSI]_2$	1×10^{-4} at 20°C	[46]

Table 2.
Ionic conductivity of ILs based GPEs.

5. Results and discussion

5.1 Surface morphology and crystallinity of IL based GPEs

The effect on surface morphology and degree of crystallinity of PEO based SPE with the addition of IL is reported in literature. Gupta et al. [47] studied that, by using the phosphonium based IL (Trihexyltetradecylphosphonium TFSI) in polymer electrolyte (PEO + 20 wt% LiTFSI), increment in amorphous region was obtained (**Figure 6**). The surface morphology of PEO based GPE is given in **Figure 6**. **Figure 6(a)** shows the crystalline region of polymer PEO, when IL (up to 20 wt%) was incorporated into PEO-LiTFSI system, smoother morphology was observed which resulted the amorphous nature of GPE. Singh et al. [48] also reported that the suppression in rough surface nature of polymer electrolyte (PEO + 20 wt% LiTFSI) was observed when BMIMTFSI IL was added into it. The smoother surface morphology of polymer electrolyte was observed due to the plasticization effect of ionic liquid which reduced the interaction between the polymer chain and made it more flexible.

From the XRD analysis, crystallinity of GPEs is also reported in many studies. Gupta et al. [47] reported the variation in crystallinity of SPE with the addition of phosphonium based IL They showed the semi-crystalline nature of polymer PEO. When LiTFSI salt and IL were added into the PEO, broadening of halo region and FDHM of polymer electrolytes were noticed which resulted the decrement in crystalline region or enhancement in amorphous region of polymer electrolyte (**Figure 7(A)**).

Singh et al. [48] reported the effect of BMIMTFSI IL on the crystallinity of SPE (PEO + 20 wt% LiTFSI). They showed that with the addition of IL into SPE, halo region was increased substantially and relative intensities of the crystalline peaks reduced. This proved the enhancement in amorphous phase in polymer electrolyte (**Figure 7(B)**). Therefore, presence of IL in polymer electrolyte improves its amorphous region which is desirable for the conduction of Li^+ ions, since, conduction in polymer electrolytes occurs only in this region.

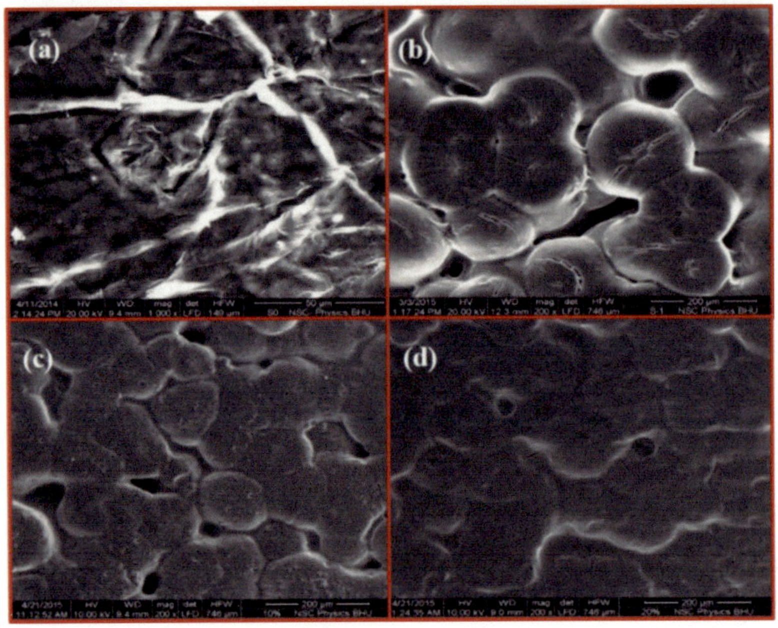

Figure 6.
SEM image for (a) pure PEO and PEO + 20 wt% LiTFSI + Xwt% IL (b) X = 0, (c) X = 10, (d) X = 20.

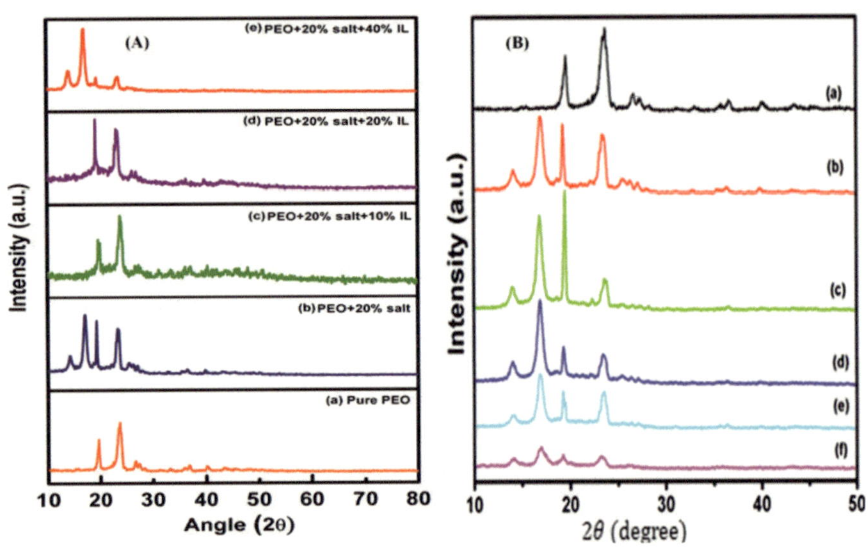

Figure 7.
XRD pattern of (A) phosphonium IL based polymer electrolytes, PEO + 20wt.%LiTFSI + x% IL (x = 0, 10, 20, 40); and (B) polymer electrolytes (a) PEO with PEO + 20wt.%LiTFSI + xwt.% BMIMTFSI (b) x = 0, (c) x = 5, (d) x = 10, (e) x = 15 and (f) x = 20.

5.2 Ionic and Li^+ ions conductivity of IL based GPEs

In batteries, electrolyte plays the major role to transport the ions between the two electrodes. So, conductivity of electrolyte is very important parameter for Li battery. It is well known that the IL acts as a plasticizer and its presence in polymer electrolyte enhances the ionic conductivity. IL also provides free charge carriers and therefore helps to promote the ionic conductivity of polymer electrolyte.

The increment in the conductivity of polymer electrolyte with IL concentration is also reported by Gupta et al. [49] and Balo et al. [36] which are depicted in **Figure 8(a)** and **(b)** respectively. Gupta et al. [49] showed that ionic conductivity of polymer electrolyte (PEO + 20 wt% LiFSI) increases with IL ($PYR_{13}FSI$) concentration.

Whereas, Balo et al. [36] synthesized the PEO based polymer electrolyte (PE O + 20wt.%LiFSI + xwt.%EMIMFSI; for $0 \leq x \leq 15$) and mentioned the use of optimized concentration of IL in polymer electrolyte. They found similar increasing trend of conductivity of polymer electrolyte with IL, but after certain concentration, it showed decreasing trend (**Figure 8(b)**). It happened because with the addition of IL into polymer electrolyte, large number of FSI^- anions was available which started to interact with Li^+ ions present in the polymer electrolyte. Therefore, they formed ion pairs instead of participating in interaction with ether oxygen of PEO, due to which conductivity of electrolyte was reduced.

5.3 Electrochemical stability of IL based GPEs

Electrochemical stability of electrolyte is an important parameter as it decides the performance of the battery in working voltage range. Electrochemical stability of GPEs should be high so that it can be used in high voltage Li batteries. Many studies have been carried out on the electrochemical performance of PEO and IL

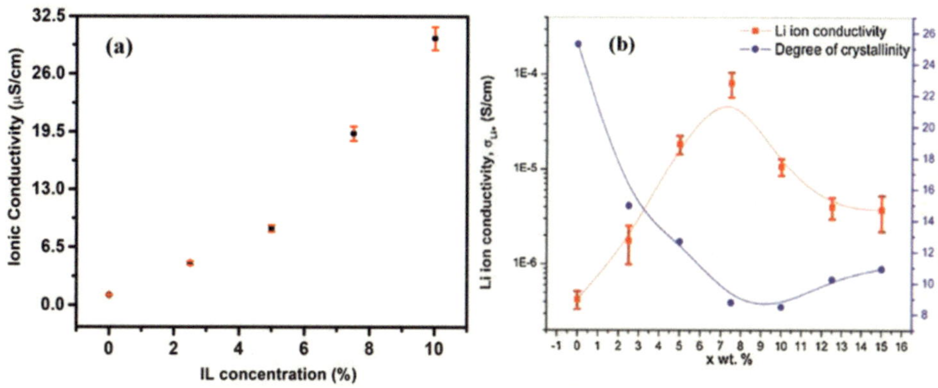

Figure 8.
Variation in (a) ionic conductivity as well as (b) Li^+ ion conductivity and degree of crystallinity of polymer electrolyte with IL concentration.

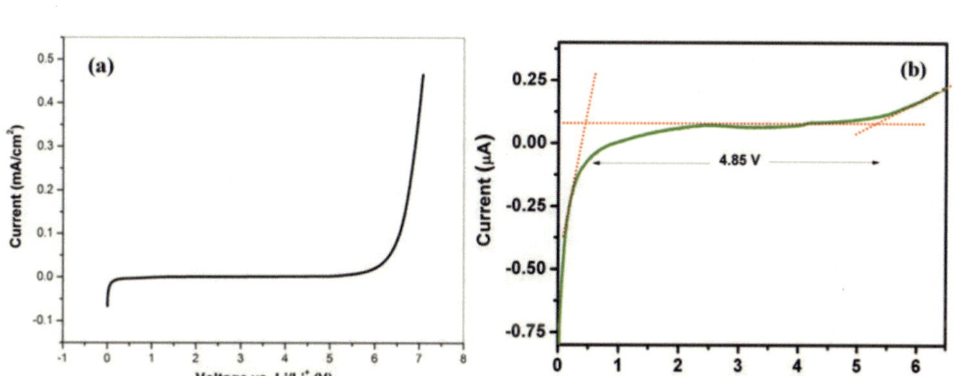

Figure 9.
Electrochemical stability of (a) imidazolium (EMIMFSI) and (b) pyrrolidinium ($PYR_{13}FSI$) IL based GPEs.

based GPEs. Singh et al. [48] synthesized the GPE, PEO + 20 wt% LITFSI + xwt% BMIMTFSI (x = 5, 10, 15, 20) for Li battery and the electrochemical stability of 20 wt% IL containing GPE was reported ~4 V vs. Li/Li^+.

Balo et al. used the same polymer system with EMIMFSI and EMIMTFSI ILs and found that the GPE, PEO + 20 wt% LiTFSI + 10 wt% EMIMFSI was electrochemically stable up to ~6.4 V vs. Li/Li^+ (**Figure 9(a)**) [50]. In another study they showed that the polymer electrolyte, PEO + 20 wt% LiTFSI + 10 wt% EMIMTFSI, was stable upto ~5.1 V vs. Li/Li^+ [42]. Also, Gupta et al. [49] reported the electrochemical stability of pyrrolidinium IL based GPE (PEO + 20% LiFSI + 10% $PYR_{13}FSI$) which was ~4.8 V vs. Li/Li^+ (**Figure 9(b)**). Therefore, it can be concluded that IL based GPEs show enough electrochemical stability to be used in high voltage Li batteries application.

6. Application of IL based GPEs in Lithium batteries

Because of having several required properties, IL based polymer electrolytes are frequently used in many application as in supercapacitors, batteries and fuel cell etc. Particularly, in Li batteries, they are gaining much attention due to their high energy density, flexibility and safety. In recent years, almost all the electronic equipment are being run by polymer batteries as laptop, mobile phones, power banks, portable media players etc. Many reports include the application of GPEs in Li batteries. As one of the main advantage of GPE is that it forms stable solid electrolyte interface (SEI) passive layer between the electrode-electrolyte and provides higher cyclic stability to Li battery. Battery performance of PEO based GPEs have already been studied in literature. Gupta et al. [51] synthesized the GPE, (PEO + 20 wt% LiTFSI + 30 wt% 1-butyl-3-methyl pyridinium TFSI), and reported its performance in Li battery in ($Li/LiFePO_4$) configuration. They obtained maximum discharge capacity ~160 mAh/g and 99% Coulombic efficiency upto 35 cycles at 40°C (**Figure 10(a)** and **(b)**). In another report, they used the pyrrolidinium based IL in polymer system, PEO + 20% LiFSI + 10% $PYR_{13}FSI$, with graphene oxide coated $LiFePO_4$ cathode and obtained maximum discharge capacity ~163 mAh/g at C/10 rate at room temperature (RT) (**Figure 10(c)**) [49]. It was the result of high conductivity of IL, $PYR_{13}FSI$ and LiFSI salt as well as high electronic conductivity and large surface area of graphene oxide (GO) which enhanced the electron transfer rate and hence capacity of Li battery. This Li battery provided almost constant capacity and Coulombic efficiency upto 100 cycles (**Figure 10(d)**). Balo et al. [36] reported the same system ($Li/LiFePO_4$) using imidazolium IL based polymer electrolyte PEO + 20 wt.% LiFSI + 7.5wt.%EMIMFSI at RT.

They found maximum discharge capacity ~143 mAh/g at C/20 rate which decreased upto 130 mAh/g at C/10 and further reduced upto 20 mAh/g at 2C discharge rate. This reduction of discharge capacity was reported due to the increase of electrolyte ohmic drop and limited Li^+ ion diffusion in composite cathode. The above polymer systems have been also tested with high voltage and capacity cathode materials such as $LiNi_xMn_yCo_zO_2$ (NMC) and $LiNi_xCo_yAl_zO_2$ (NCA). These electrolytes are electrochemically stable even at high voltage which deliver high capacity and cyclic stability to the Li battery. Gupta et al. [52] used the phosphonium based IL (trihexyltetradecylphosphonium bis TFSI) in PEO-LiTFSI polymer system.

They fabricated the Li cell ($Li/LiNi_{0.6}Mn_{0.2}Co_{0.2}O_2$) and obtained maximum discharge capacity ~148 mAh/g with 95% Coulombic efficiency upto 150th cycle at C/10 rate in the voltage range of 2.4–4.2 V (**Figure 11(a, b)**). The impedance of the Li cell was also evaluated with cycling (inset of **Figure 11(b)**). It showed the slight increment in the interfacial resistance value and hence, resulted very small capacity fading of Li cell (**Figure 11(b)**). Balo et al. [50] used imidazolium IL (EMIMFSI)

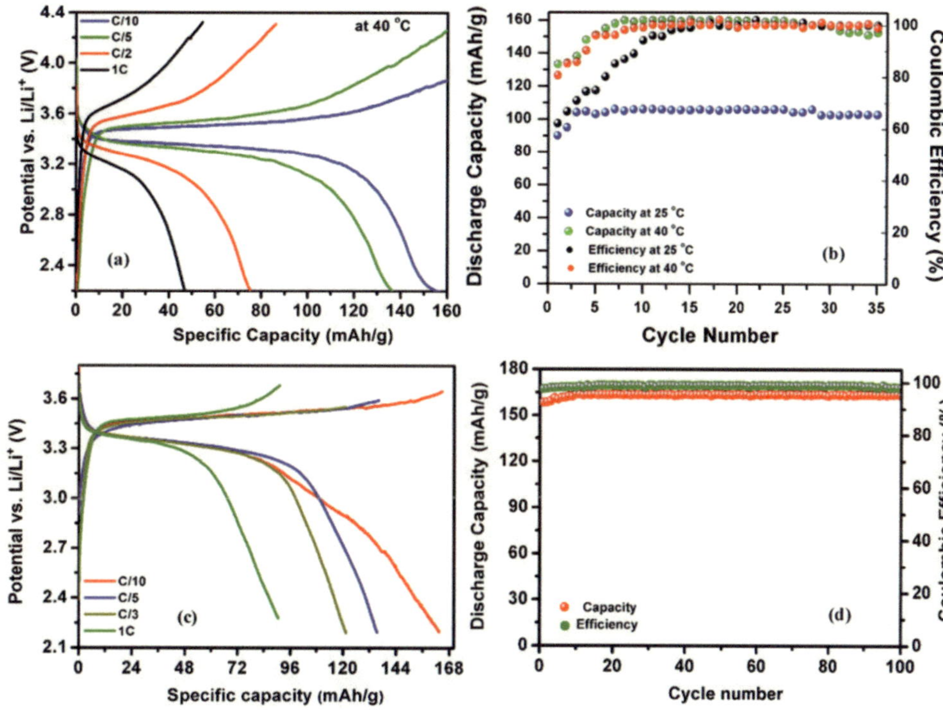

Figure 10.
Electrochemical performance of (a, b) 1-butyl-3-methyl pyridinium TFSI and (c, d) N-propyl-N-methylpyrrolidinium-FSI IL based polymer electrolytes with Li/$LiFePO_4$ and graphene oxide coated $LiFePO_4$ cathode respectively.

based GPE (PEO + 20 wt% LiTFSI + 10 wt% EMIMFSI) in Li/NCA cell. They observed the discharge capacity ~175 mAh/g at C/10 current rate which remained almost stable throughout cycling (**Figure 11(c, d)**) and only 0.05% of total capacity was lost during 200 cycles (inset of **Figure 11(d)**). The use of Li metal electrode in Li batteries are in demand due to its higher energy density and capacity (3862 mAh/g), but it could not be frequently used in application purpose because of the formation of Li dendrites. This Li dendrite is formed due to the deposition of Li^+ ions on the Li metal surface during cycling which starts to grow and causes short circuiting and results low cyclic stability. Therefore, in order to obtain high capacity and safer Li battery, suppression of dendrite growth is important. It was reported that the dendrite growth becomes faster with liquid solvents. But the use of GPEs in Li battery is able to suppress its growth because of having mechanical stability. Therefore, the use of GPEs provides safety and cyclic stability to Li battery. The electrochemical stability of GPEs with Li electrode is reported in literature. Wang et al. [53] reported the combination of the use of LAGP-PEO (LiTFSI) composite solid electrolyte and the modification of Li anode with PEO_{500000} (LiTFSI) in Li/$LiMn_{0.8}Fe_{0.2}PO_4$ battery. They obtained that the use of both can effectively prevent the Li dendrite growth. Kim et al. used three different ILs (BMITFSI, $BMIBF_4$ and $BMICF_3SO_3$) in polymer system (PEO-LiTFSI) and reported that the Polymer electrolyte with BMITFSI IL resulted low and stable interfacial resistance or dendrite growth on lithium metal. Balo et al. [50] examined the performance of EMIMFSI IL in PEO-LiTFSI system. They found the stable and uniform formation of Li dendrite between lithium and GPE during cycling (**Figure 12(a)**). They also evaluated the interfacial resistance of this passive layer and observed that except the initial few cycles almost stable interfacial resistance 380 Ω/cm^2 was obtained throughout the cycling (**Figure 12(b)**).

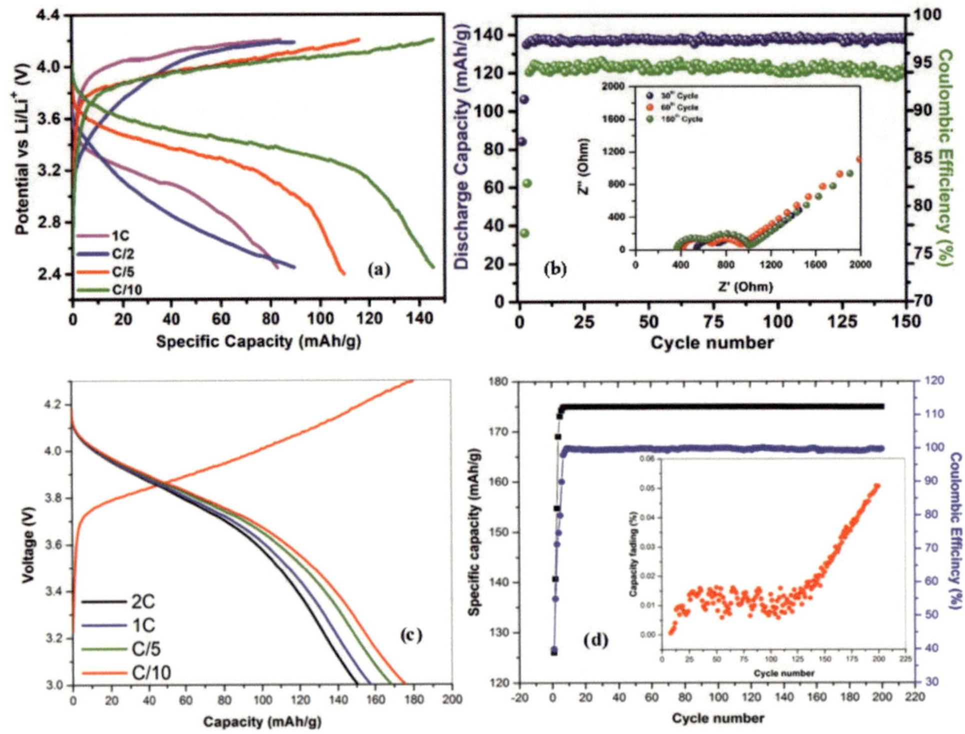

Figure 11.
Discharge capacity, efficiency (at C/10 rate) and capacity fading of the Li cell (a, b) Li/(PEO + 20wt%LiTFSI + 20wt% trihexyltetradecylphosphonium TFSI)/$LiNi_{0.6}Mn_{0.2}Co_{0.2}O_2$ and (c, d) Li/(PEO + 20 wt% LiTFSI + 10 wt% EMIMFSI)/NCA.

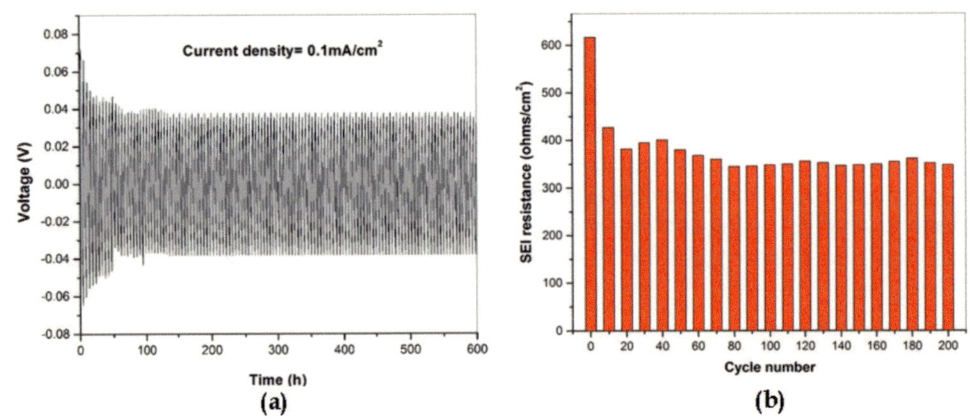

Figure 12.
(a) Voltage vs. time profile of lithium deposition and (b) evolution of interfacial resistances during cycling using GPE (PEO + 20wt% LiTFSI + 10wt% EMIMFSI).

Other reports on the electrochemical performance of Li batteries using PEO based GPEs are also tabulated in **Table 3**. All these analysis shows that the use of GPEs in Li battery maintains the cyclability and electrochemical stability of the Li battery much more compared to liquid solvents.

Therefore, from the above discussions it can be concluded that the IL based GPEs not only provide good ionic conductivity, flexibility and mechanical stability but also act as a potential candidate in order to enhance the capacity, cyclicity and safety to Li battery.

Polymer electrolytes	Li battery	C-rate	Capacity (mAh/g)	References
$PEO_{20}LiTFSI[Pyr13TFSI]_{1.27}$	$Li/LiFePO_4$ at RT	C/10	115 at 20 cycles	[54]
$PEO_{20}LiTFSI2[Pyr14TFSI]_4$	Li/NMC at 40°C	C/10	160 at 100th cycle	[55]
$PEO_{20}LiTFSI[Im12TFSI]$	$Li/LiFePO_4$ at 50°C	C/5	110 at 20th cycle	[56]
$PEO_{20}LiTFSI[Pip_{1.101}TFSI]$	$Li/LiFePO_4$ at RT	C/20	120 at 35th cycle	[57]
$PEO_{20}LiTFSI[Pip_{1.101}TFSI]$	$Li/Li_4Ti_5O_{12}$ at RT	C/20	150 at 40th cycle	[57]
$PEO_{20}LiTFSI_2[Pyr14TFSI]_4$	$Li/LiFePO_4$ at 40°C	C/10	140 at 450th cycle	[58]
PEO + 20wt%LiTFSI + 20wt%ThdpTFSI	Li/NMC622 at RT	C/10	148 at 150th cycle	[52]
$PEO_{20}LiTFSI_2[Pyr14TFSI]_2$	$Li/LiFePO_4$ at 40°C	C/5	160 at 180th cycle	[59]
PEO-LiTFSI-10wt%EMIMFSI	Li/NCA at RT	C/10	175 at 200th cycle	[50]
PEO + LiFSI + 7.5 wt.% EMIMFSI	$Li/LiFePO_4$ at RT	C/20	143 at 100th cycle	[36]
PEO + LiTFSI + 12.5%EMIMTFSI	$Li/LiMn_2O_4$ at RT	C/10	120 at 10th cycle	[42]
PEO + 20%LiTFSI + 30%(1-butyl 3-methyl pyridiniumTFSI)	$Li/LiFePO_4$ at 40°C	C/10	160 at 25th cycle	[51]
PEO + 20%LiFSI + 10%PYR13FSI	$Li/GO-LiFePO_4$ at rt	C/10	163 at 100th cycle	[49]
PEO + 20wt%LiTFSI + 20 wt% BMIMTFSI	$Li/LiMn_2O_4$ at RT		140 $\mu Ah/cm^2$ at 25 cycle	[48]

Table 3.
Electrochemical performance of Li batteries using IL based GPEs.

7. Conclusion

In summary, many approaches have been proposed to enhance the conductivity of PEO based polymer electrolyte below melting temperature. Among them, ionic liquid based GPEs are considered as the most promising approach. These GPEs enhance the ionic conductivity, thermal and electrochemical stability of the polymer electrolytes. They provide better electrode-electrolyte contact, mechanical stability and safety to Li batteries. Because of having enough mechanical stability, they are able to suppress the undesirable dendrite formation and help to provide safer Li battery. Also, due to their high electrochemical stability, they can be used in high voltage and energy density batteries. In these batteries, they show good electrochemical and cyclic stability as well as offer flexibility and safety. Therefore, these IL based GPEs can be considered as a potential candidate for replacement of liquid electrolyte in Li batteries.

Acknowledgements

One of the authors (RKS) thankfully acknowledges the financial support by DST-SERB, New Delhi, BRNS-DAE to carry out the research work. H. Gupta is grateful to DST, India for providing SRF project fellowship.

Conflict of interest

The authors declare no conflict of interest.

Author details

Himani Gupta and Rajendra K. Singh*
Department of Physics, Institute of Science, Banaras Hindu University, Varanasi, India

*Address all correspondence to: rajendrasingh.bhu@gmail.com

References

[1] Kim BK, Sy S, Yu A, Zhang J. Electrochemical supercapacitors for energy storage and conversion. Handbook of Clean Energy Systems. John Wiley & Sons, Ltd.; 2015:1-25. DOI: 10.1002/9781118991978.hces112

[2] Monzer AS, Gualous H, Omar N, Mierlo JV. Batteries and supercapacitors for electric vehicles. New Generation of Electric Vehicles. InTech. 2012:135-164. DOI: 10.5772/53490

[3] Whittingham MS. Lithium batteries and cathode materials. Chemical Reviews. 2004;**104**:4271-4301. DOI: 10.1021/cr020731c

[4] Lu L, Han X, Li J, Hua J, Ouyang M. A review on the key issues for lithium-ion battery management in electric vehicles. Journal of Power Sources. 2013;**226**:272-288. DOI: 10.1016/j.jpowsour.2012.10.060

[5] Conte FV. Battery and battery management for hybrid electric vehicles: A review. e & i Elektrotechnik und Informationstechnik. 2006;**123**:424-431. DOI: 10.1007/ s00502-006-0383-6

[6] Ritchie AG. Recent developments and future prospects for lithium rechargeable batteries. Fuel and Energy Abstracts. 2002;**43**:193. DOI: 10.1016/ s0140-6701(02)85787-3

[7] Lv D, Shao Y, Lozano T, Bennett WD, Graff GL, Polzin B, et al. Failure mechanism for fast-charged lithium metal batteries with liquid electrolytes. Advanced Energy Materials. 2015;**5**:1-7. DOI: 10.1002/aenm.201400993

[8] Xu W, Wang J, Ding F, Chen X, Nasybulin E, Zhang Y, et al. Lithium metal anodes for rechargeable batteries. Energy and Environmental Science. 2014;**7**:513-537. DOI: 10.1039/ c3ee40795k

[9] Ould Ely T, Kamzabek D, Chakraborty D. Batteries safety: Recent progress and current challenges. Frontiers in Energy Research. 2019;7:1-44. DOI: 10.3389/fenrg.2019.00071

[10] Liu Z, Huang Y, Huang Y, Yang Q, Li X, Huang Z, et al. Voltage issue of aqueous rechargeable metal-ion batteries. Chemical Society Reviews. 2020;**49**:180-232. DOI: 10.1039/ c9cs00131j

[11] Wang Y, Zhong WH. Development of electrolytes towards achieving safe and high-performance energy-storage devices: A review. ChemElectroChem. 2015;**2**:22-36. DOI: 10.1002/celc.201402277

[12] Srivastava S, Schaefer JL, Yang Z, Tu Z, Archer LA. 25th anniversary article: Polymer-particle composites: Phase stability and applications in electrochemical energy storage. Advanced Materials. 2014;**26**:201-234. DOI: 10.1002/adma.201303070

[13] Aziz SB. Li+ ion conduction mechanism in poly (ε-caprolactone)-based polymer electrolyte. Iranian Polymer Journal (English Edition). 2013;**22**:877-883. DOI: 10.1007/ s13726-013-0186-7

[14] Wright P v. Electrical conductivity in ionic complexes of poly(ethylene oxide). British Polymer Journal. 1975;**7**:319-327. DOI: 10.1002/ pi.4980070505

[15] Armand M. Polymer electrolytes. Annual Review of Materials Science. 1986;**16**:245-261. DOI: 10.1002/9783527611676.ch20

[16] Cheng S, Smith DM, Li CY. How does nanoscale crystalline structure affect ion transport in solid polymer electrolytes? Macromolecules. 2014;**47**:3978-3986. DOI: 10.1021/ ma500734q

[17] Johansson P. First principles modelling of amorphous polymer electrolytes: Li+-PEO, Li+-PEI, and Li+-PES complexes. Polymer. 2001;**42**:4367-4373. DOI: 10.1016/S0032-3861(00)00731-X

[18] Geiculescu OE, Hallac BB, Rajagopal RV, Creager SE, Desmarteau DD, Borodin O, et al. The effect of low-molecular-weight poly(ethylene glycol) (PEG) plasticizers on the transport properties of lithium fluorosulfonimide ionic melt electrolytes. Journal of Physical Chemistry B. 2014;**118**:5135-5143. DOI: 10.1021/jp500826c

[19] Song JY, Wang YY, Wan CC. Review of gel-type polymer electrolytes for lithium-ion batteries. Journal of Power Sources. 1999;**77**:183-197. DOI: 10.1016/S0378-7753(98)00193-1

[20] Manuel Stephan A, Nahm KS. Review on composite polymer electrolytes for lithium batteries. Polymer. 2006;**47**:5952-5964. DOI: 10.1016/j.polymer.2006.05.069

[21] Xue Z, He D, Xie X. Poly(ethylene oxide)-based electrolytes for lithium-ion batteries. Journal of Materials Chemistry A. 2015;**3**:19218-19253. DOI: 10.1039/c5ta03471j

[22] Long L, Wang S, Xiao M, Meng Y. Polymer electrolytes for lithium polymer batteries. Journal of Materials Chemistry A. 2016;**4**:10038-10039. DOI: 10.1039/c6ta02621d

[23] Chen R, Qu W, Guo X, Li L, Wu F. The pursuit of solid-state electrolytes for lithium batteries: From comprehensive insight to emerging horizons. Materials Horizons. 2016;**3**:487-516. DOI: 10.1039/C6MH00218H

[24] Edman L, Doeff MM, Ferry A, Kerr J, de Jonghe LC. Transport properties of the solid polymer electrolyte system P(EO)nLiTFSI. Journal of Physical Chemistry B. 2000;**104**:3476-3480. DOI: 10.1021/jp993897z

[25] Rivas BL, Maureira AE, Mondaca MA. Aminodiacetic water-soluble polymer-metal ion interactions. European Polymer Journal. 2008;**44**:2330-2338. DOI: 10.1016/j.eurpolymj.2008.05.001

[26] Stephan AM, Kumar TP, Kulandainathan MA, Lakshmi NA. Chitin-incorporated poly(ethylene oxide)-based nanocomposite electrolytes for lithium batteries. Journal of Physical Chemistry B. 2009;**113**:1963-1971. DOI: 10.1021/jp808640j

[27] Honary S, Orafai H. The effect of different plasticizer molecular weights and concentrations on mechanical and thermomechanical properties of free films. Drug Development and Industrial Pharmacy. 2002;**28**:711-715. DOI: 10.1081/DDC-120003863

[28] Mohapatra SR, Thakur AK, Choudhary RNP. Effect of nanoscopic confinement on improvement in ion conduction and stability properties of an intercalated polymer nanocomposite electrolyte for energy storage applications. Journal of Power Sources. 2009;**191**:601-613. DOI: 10.1016/j.jpowsour.2009.01.100

[29] Mulmi S, Park CH, Kim HK, Lee CH, Park HB, Lee YM. Surfactant-assisted polymer electrolyte nanocomposite membranes for fuel cells. Journal of Membrane Science. 2009;**344**:288-296. DOI: 10.1016/j.memsci.2009.08.028

[30] Agrawal RC, Pandey GP. Solid polymer electrolytes: Materials designing and all-solid-state battery applications: An overview. Journal of Physics D: Applied Physics. 2008;**41**:223001(1-18). DOI: 10.1088/0022-3727/41/22/223001

[31] Lu J, Yan F, Texter J. Advanced applications of ionic liquids in polymer science. Progress in Polymer Science (Oxford). 2009;**34**:431-448. DOI: 10.1016/j.progpolymsci.2008.12.001

[32] Amde M, Liu JF, Pang L. Environmental application, fate, effects, and concerns of ionic liquids: A review. Environmental Science and Technology. 2015;**49**:12611-12627. DOI: 10.1021/acs.est.5b03123

[33] Huddleston JG, Visser AE, Reichert WM, Willauer HD, Broker GA, Rogers RD. Characterization and comparison of hydrophilic and hydrophobic room temperature ionic liquids incorporating the imidazolium cation. Green Chemistry. 2001;**3**:156-164. DOI: 10.1039/b103275p

[34] Zhang S, Sun N, He X, Lu X, Zhang X. Physical properties of ionic liquids: Database and evaluation. Journal of Physical and Chemical Reference Data. 2006;**35**:1475-1517. DOI: 10.1063/1.2204959

[35] Abitelli E, Ferrari S, Quartarone E, Mustarelli P, Magistris A, Fagnoni M, et al. Polyethylene oxide electrolyte membranes with pyrrolidinium-based ionic liquids. Electrochimica Acta. 2010;**55**:5478-5484. DOI: 10.1016/j.electacta.2010.04.099

[36] Balo L, Gupta H, Singh SK, Singh VK, Kataria S, Singh RK. Performance of EMIMFSI ionic liquid based gel polymer electrolyte in rechargeable lithium metal batteries. Journal of Industrial and Engineering Chemistry. 2018;**65**:137-145. DOI: 10.1016/j.jiec.2018.04.022

[37] de Vries H, Jeong S, Passerini S. Ternary polymer electrolytes incorporating pyrrolidinium-imide ionic liquids. RSC Advances. 2015;**5**:13598-13606. DOI: 10.1039/c4ra16070c

[38] Shalu BL, Gupta H, Singh VK, Singh RK. Mixed anion effect on the ionic transport behavior, complexation and various physicochemical properties of ionic liquid based polymer gel electrolyte membranes. RSC Advances. 2016;**6**:73028-73039. DOI: 10.1039/c6ra10340e

[39] Nakagawa H, Izuchi S, Kuwana K, Nukuda T, Aihara Y. Liquid and polymer gel electrolytes for lithium batteries composed of room-temperature molten salt doped by lithium salt. Journal of the Electrochemical Society. 2003;**150**:A695. DOI: 10.1149/1.1568939

[40] Pappenfus TM, Henderson WA, Owens BB, Mann KR, Smyrl WH. Complexes of lithium imide salts with tetraglyme and their polyelectrolyte composite materials. Journal of the Electrochemical Society. 2004;**151**:A209. DOI: 10.1149/1.1635384

[41] Wu L. Chloride ion conducting polymer electrolytes based on cross-linked PMMA-PP14Cl-PP14TFSI ion gels for chloride ion batteries. International Journal of Electrochemical Science. 2019;**14**:2414-2421. DOI: 10.20964/2019.03.49

[42] Balo L, Shalu GH, Kumar Singh V, Kumar Singh R. Flexible gel polymer electrolyte based on ionic liquid EMIMTFSI for rechargeable battery application. Electrochimica Acta. 2017;**230**:123-131. DOI: 10.1016/j.electacta.2017.01.177

[43] Shin JH, Henderson WA, Passerini S. Ionic liquids to the rescue? Overcoming the ionic conductivity limitations of polymer electrolytes. Electrochemistry Communications. 2003;**5**:1016-1020. DOI: 10.1016/j.elecom.2003.09.017

[44] Singh VK, Shalu CSK, Singh RK. Development of ionic liquid mediated novel polymer electrolyte membranes for application in Na-ion batteries. RSC

Advances. 2016;**6**:40199-40210. DOI: 10.1039/c6ra06047a

[45] Singh VK, Singh SK, Gupta H, Shalu BL, Tripathi AK, et al. Electrochemical investigations of $Na_{0.7}CoO_2$ cathode with PEO-NaTFSI-BMIMTFSI electrolyte as promising material for Na-rechargeable battery. Journal of Solid State Electrochemistry. 2018;**2**:1-11. DOI: 10.1007/s10008-018-3891-5

[46] Rupp B, Schmuck M, Balducci A, Winter M, Kern W. Polymer electrolyte for lithium batteries based on photochemically crosslinked poly(ethylene oxide) and ionic liquid. European Polymer Journal. 2008;**44**:2986-2990. DOI: 10.1016/j.eurpolymj.2008.06.022

[47] Gupta H, Shalu BL, Singh VK, Chaurasia SK, Singh RK. Effect of phosphonium based ionic liquid on structural, electrochemical and thermal behaviour of polymer poly(ethylene oxide) containing salt lithium bis(trifluoromethylsulfonyl)imide. RSC Advances. 2016;**6**:87878-87887. DOI: 10.1039/c6ra20393k

[48] Singh VK, Shalu BL, Gupta H, Singh SK, Singh RK. Solid polymer electrolytes based on Li+/ionic liquid for lithium secondary batteries. Journal of Solid State Electrochemistry. 2017;**21**:1713-1723. DOI: 10.1007/s10008-017-3529-z

[49] Gupta H, Kataria S, Balo L, Singh VK, Singh SK, Tripathi AK, et al. Electrochemical study of ionic liquid based polymer electrolyte with graphene oxide coated $LiFePO_4$ cathode for Li battery. Solid State Ionics. 2018;**320**:186-192. DOI: 10.1016/j.ssi.2018.03.008

[50] Balo L, Gupta H, Singh SK, Singh VK, Tripathi AK, Srivastava N, et al. Development of gel polymer electrolyte based on LiTFSI and EMIMFSI for application in rechargeable lithium metal battery with GO-LFP and NCA cathodes. Journal of Solid State Electrochemistry. 2019;**23**:2507-2518. DOI: 10.1007/s10008-019-04321-6

[51] Gupta H, Shalu BL, Singh VK, Singh SK, Tripathi AK, et al. Effect of temperature on electrochemical performance of ionic liquid based polymer electrolyte with Li/$LiFePO_4$ electrodes. Solid State Ionics. 2017;**309**:192-199. DOI: 10.1016/j.ssi.2017.07.019

[52] Gupta H, Singh SK, Singh VK, Tripathi AK, Srivastava N, Tiwari RK, et al. Development of polymer electrolyte and cathode material for Li-batteries. Journal of the Electrochemical Society. 2019;**166**:A5187-A5192. DOI: 10.1149/2.0331903jes

[53] Wang C, Yang Y, Liu X, Zhong H, Xu H, Xu Z, et al. Suppression of lithium dendrite formation by using LAGP-PEO (LiTFSI) composite solid electrolyte and lithium metal anode modified by PEO (LiTFSI) in all-solid-state lithium batteries. ACS Applied Materials and Interfaces. 2017;**9**:13694-13702. DOI: 10.1021/acsami.7b00336

[54] An Y, Cheng X, Zuo P, Liao L, Yin G. Improved properties of polymer electrolyte by ionic liquid PP1.3TFSI for secondary lithium ion battery. Journal of Solid State Electrochemistry. 2012;**16**:383-389. DOI: 10.1007/s10008-011-1340-9

[55] Wetjen M, Kim GT, Joost M, Appetecchi GB, Winter M, Passerini S. Thermal and electrochemical properties of PEO-LiTFSI-Pyr 14TFSI-based composite cathodes, incorporating 4V-class cathode active materials. Journal of Power Sources. 2014;**246**:846-857. DOI: 10.1016/j.jpowsour.2013.08.037

[56] Zhu C, Cheng H, Yang Y. Electrochemical characterization of two

types of PEO-based polymer electrolytes with room-temperature ionic liquids. Journal of the Electrochemical Society. 2008;**155**:A569. DOI: 10.1149/1.2931523

[57] An Y, Cheng X, Zuo P, Liao L, Yin G. The effects of functional ionic liquid on properties of solid polymer electrolyte. Materials Chemistry and Physics. 2011;**128**:250-255. DOI: 10.1016/j.matchemphys.2011.03.007

[58] Kim GT, Appetecchi GB, Carewska M, Joost M, Balducci A, Winter M, et al. UV cross-linked, lithium-conducting ternary polymer electrolytes containing ionic liquids. Journal of Power Sources. 2010;**195**:6130-6137. DOI: 10.1016/j.jpowsour.2009.10.079

[59] Appetecchi GB, Kim GT, Montanino M, Alessandrini F, Passerini S. Room temperature lithium polymer batteries based on ionic liquids. Journal of Power Sources. 2011;**196**:6703-6709. DOI: 10.1016/j.jpowsour.2010.11.070

Section 3

Bio-Electrochemical Process

Chapter 6

Bioelectrochemical Processes in Industrial Biotechnology

Venko Beschkov and Elena Razkazova-Velkova

Abstract

Industrial fermentation and biological wastewater treatment are usually based on redox processes taking place in living cells and on enzyme processes. The practical application of these redox processes is usually associated with electricity generation in microbial fuel cells or process enhancement in microbial electrolysis cells. The microbial fuel cell approach leads to straightforward wastewater treatment with less energy demand. Additional advantages of these processes are the direct removal of various pollutants and the avoidance of addition of chemical agents with the resulting waste products of treatment as it is familiar with the traditional chemical methods. Another option for the use of bioelectrochemical processes in practice is the approach of microbial electrolysis cells. The application of electric field on fermentation or microbial wastewater treatment processes might result in different aspects: either in purely electrochemical processes on the electrodes or in different types of bioelectrochemical stimulation of enzyme activity in the living cells. These applications are associated with the combination of enzyme activity with electrochemical processes to produce or remove certain compounds rapidly at high concentrations with no additions of other chemicals. In the present chapter, both approaches (microbial fuel cells and microbial electrolysis cells) are presented and discussed. Some practical applications and experimental examples of such bioelectrochemical redox processes stimulated by constant electric field are demonstrated.

Keywords: redox systems, microbial fuel cells, microbial electrolysis cells, bioelectrochemical oxidation, bioelectrochemical reduction

1. Introduction

The depletion of the traditional energy resources together with the environmental problems caused by the excessive use of fossil fuels and the resulting emissions of greenhouse gases have prompted humanity to replace fossil fuels, at least partially, by renewable energy sources. Besides the well-known hydropower, wind power and solar energy as well as utilization of biomass, there is another option to remedy the problem of energy demand and the resulting pollution. It is based on bioelectrochemical processes.

During the last decades the concept of microbial fuel cells (MFC) is a subject of significant scientific interest [1–4]. This idea seems to be very attractive, because it offers double benefit: first, to generate electricity without air pollution and second, to carry out wastewater treatment to clean water ponds with considerably reduced energy consumption [3, 4]. Although the electric power density for these devices is rather modest, the generated energy might reduce the energy consumption of a wastewater treatment plant compared to the traditional ones and to open the way to further improvements.

Another option for bioelectrochemical applications in wastewater treatment and industrial biotechnology is the stimulation of microbial redox processes by electric field.

The present work proposes an overview on the principles, achievements and future trends in these two fields of scientific and practical activity: energy generation by microbial fuel cells and enhanced processes in biotechnology and wastewater treatment by microbial electrolysis cells (MEC).

2. Microbial fuel cells

The principle of operation for microbial fuel cells is similar to the one for the traditional fuel cells, cf. **Figure 1**. There are fuels, i.e. reducers, and oxidant, usually oxygen or air. A typical feature of the microbial fuel cells is that, they operate at ambient temperature with a little external energy input to maintain the redox process. The compartment the microbes are placed in depends on their role in the overall process.

In general, the electrochemical reactions taking place on the electrodes are:

On the cathode:

$$O_2 + 4H^+ + 4e^- \rightarrow 2H_2O \tag{1}$$

On the anode:

$$\text{Reducers (represented as chemical oxygen demand, COD)} - 4e^- - 4H^+ \rightarrow \text{oxidation products.} \tag{2}$$

Of course, the electron acceptor could be different, e.g. sulfate, nitrate, etc. Thus the beneficial effect could be multiple – cleaning wastewater from excessive COD and removing nitrate from surface or industrial water, for example.

The protons necessary for the cathode reaction are transferred from the anodic space to the cathodic one through a separation proton exchanging membrane (PEM).

Electromotive force ΔU is generated as a result of the two electrode reactions. Any reducer can be used or treated by this method: starting with heavy metal cations to organic substances.

Typical characteristics of the fuel cell are the polarization curves of the dependence of the fuel cell electromotive force on the current density. An example of such polarization curve for any fuel cell is shown in **Figure 2**.

Besides the linear Ohmic range depending on the fuel cell conductivity, there is a sharp drop in voltage at low current densities, associated with the overpotential for overcoming energy barrier (reaction rate losses) [5]. At higher current densities the non-linear course of the curve is due to mass transfer limitations [5]. The polarization curves give information about the difficulties to be overcome. Moreover it is evident, that the losses are bigger for the cathode processes compared to the anodic space and the membrane resistance.

When the target process is removal of reducers, the conditions in the anode space of the MFC must be anaerobic to avoid competitive oxidation of the substrates in the bulk. This is one of the reasons why the MFC efficiencies are rather

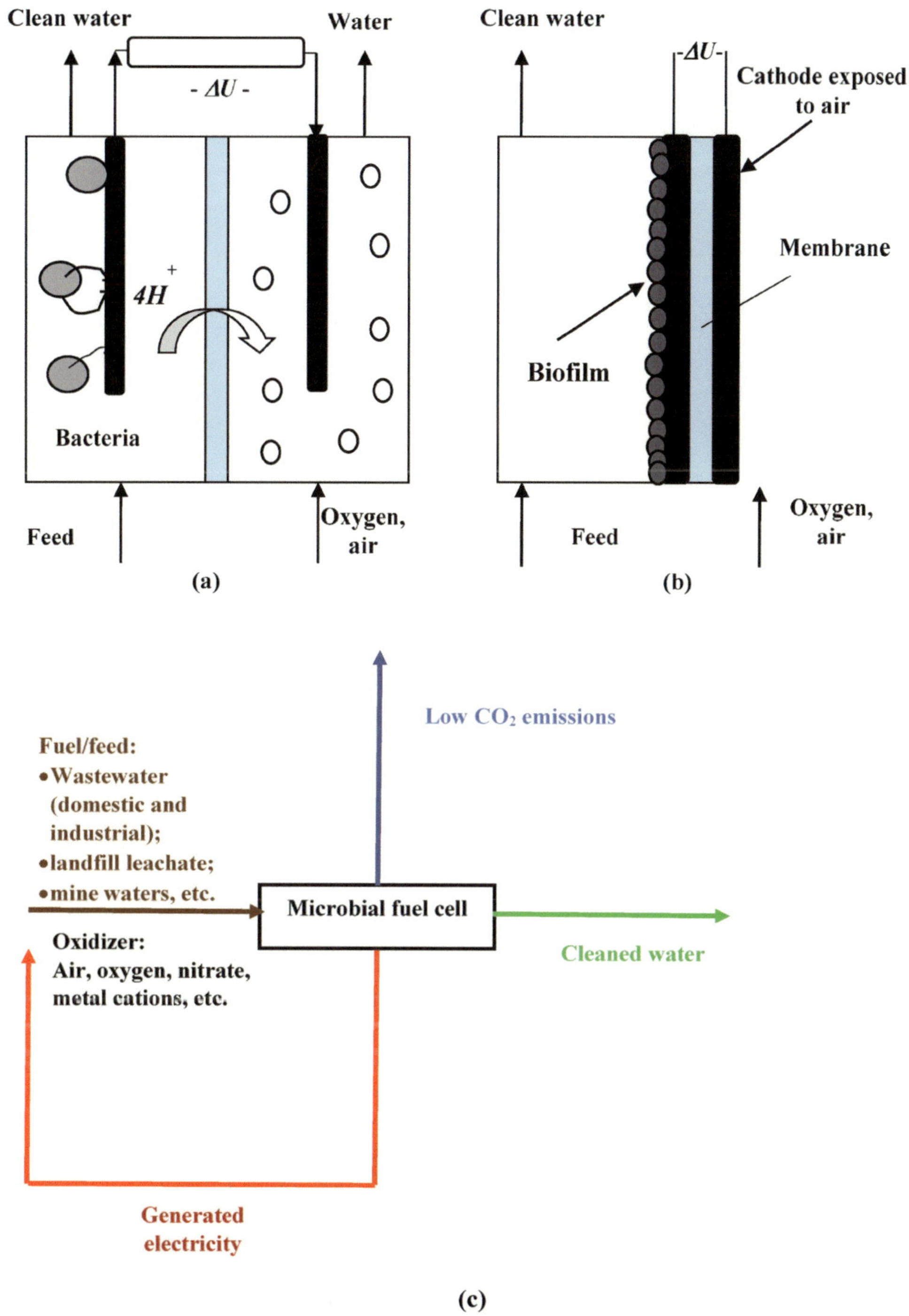

Figure 1.
A general drawing of a microbial fuel cell with oxygen as electron acceptor. Two-chamber set with solution-fed cathode (a); one-chamber set (b). Sketch of the MFC application mode (c).

low. Therefore enhancement of the electron transfer to the electrode is required. It is accomplished by different means: self-produced mediators by the living cells, artificial ones and by direct electron transfer to the anode [6, 7]. It is important to select bacteria that can directly transport electrons outside the cells, so-called "exoelectrogens", cf. Logan et al. [8].

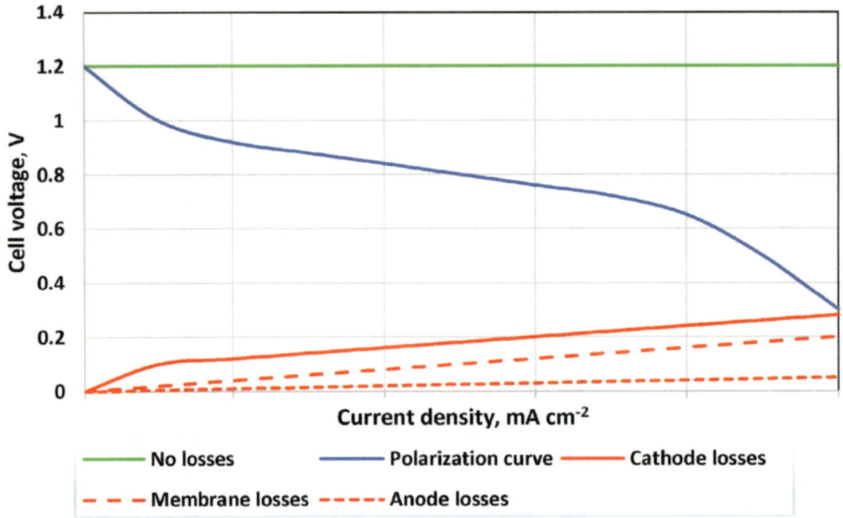

Figure 2.
Typical polarization curve of fuel cell operating in gaseous phase.

Self-produced mediators, e.g. pyocyanin can shuttle electrons and produce electricity. This method was first proposed as a mechanism for electron transfer to Fe^{3+} by Rabaey *et al*. in 2005 [9]. The major advantage of self-produced electron mediators is the long-range interaction between the bacteria and the anode. The microorganisms do not need to be in direct contact with the electrode for the transfer of electrons to occur.

Artificial mediators can cross cell membranes and accept electrons, leaving the cell in a reduced form to transfer the electrons to the electrode [10]. Examples of such artificial mediators are the dye neutral red, iron chelates, various phenazines, 4-napthoquinone, and thionine [9]. There are studies proposing other artificial mediators for electron transfer [11, 12].

In general, there are two basic types of microbial fuel cells depending on the state of the microorganisms: with free microbial cells or with ones attached to solid support and to one of the electrodes as well. The latter approach is preferable for continuous operation also for easier electron transfer toward the electrodes.

Direct electron transfer can occur through direct contact between the microorganisms and the electrode or with the use of nanowires. Direct contact requires that the organisms have membrane bound electron transport protein relays, such as cytochromes, to facilitate the transfer of electrons out of the cell [13]. However, this transfer allows for only one layer of bacteria in direct contact with the electrode. There are bacteria with appendages called nanowires. These appendages that are supposed to carry electrons from the bacterial cell to the surface of the anode. These appendages allow for multiple layers of bacteria on the anode to transfer electrons as well as interspecies transfer and transfer from the bulk liquid to the anode [7]. Such bacteria are *Shewanella oneidensis, Synechocystis* strain PCC 6803, *Geobacter sulfurreducens,* etc. [3, 14]. Nanowires can serve as cell connections of bacteria in a biofilm and can facilitate the transfer of electrons from the outer layers of a biofilm to the electrode [15]. The lack of nanowires may result in decrease of current production by 70% [16].

All these features of electron transfer to the anode accompanied by competitive bulk reactions and fermentation processes inside the bacteria lead to low current and power efficiencies. Moreover, certain overpotentials due to limitations of electron transfer described above are possible.

Different substrates are tested for treatment by microbial fuel cells: lignocellulose materials and waste, landfill leachate, manure, low-grade carbohydrates, starch, domestic wastewater, etc. [17]. There are studies on biodegradation of hydrocarbons, including aromatic, polycyclic and heterocyclic ones by MFC [18, 19].

In some studies sulfate [20–22] or nitrate [23–26] was used as electron acceptor in the cathodic compartment, thus cleaning the water from these pollutants.

There are papers studying the reduction of sulfate to sulfide to remove the cations of heavy metals (copper, lead, zinc, etc.) as insoluble sulfide from mining drainage water [27, 28].

Proper cell design, electrodes, substrates, proton exchange membranes and bacterial species forming biofilms on the electrode are very important to microbial fuel cells [3, 7, 14, 16]. MFC requires selection of appropriate electrode materials with high electric conductivity and good adhesive properties at the same time [29–31].

Another important goal of the electrode composition is the electro-catalytic activity, attained by doping with target catalysts. The energy generation of MFC is influenced by many factors, including the type of electrode material [30–32]. Studies on this subject were made by Hubenova et al. [12] proposing styryl-quinolinum dye as molecular electrocatalyst and Mitov et al. [33], who studied nickel-doped cathode. Another study was dedicated to nano-modified NiFe-and NiFeP-carbon felt as anode electrocatalysts in yeast-biofuel cell [34]. There are also works on the use of Pt-coated carbon felt, cf. Park and Zeikus [30] and graphite electrode in a brush-like shape, cf. Logan et al. [35].

As can be seen in **Figure 2**, the cathode losses in cell voltage are the biggest. There are some efforts to minimize, if not to avoid, those losses. One reason is the limited mass transfer rate of oxygen in dual chamber fuel cell in liquid phase, cf. **Figure 1**. Therefore, the direct contact of oxygen molecules with the cathode is proposed in a single chamber fuel cell, cf. Liu et al. [36, 37]. Another approach is to replace oxygen by ferric or copper cations in aqueous solution, proposed by ter Heijne [38]. Power densities reaching 2 W m^{-2} were attained.

There are different microbial strains used for MFC. The most frequently used ones are from the genera *Geobacter* [16], *Shewanella* [3, 30], *Desulfovibrio* [39, 40], *Pseudomonas* [8, 41, 42], as well as *Clostridium* [43, 44], etc.

2.1 Microbial fuel cells for wastewater treatment (COD removal)

The most widespread application of microbial fuel cells is wastewater treatment, either of municipal wastewater or industrial ones. The pollutants, represented as chemical oxygen demand (COD) are used as a fuel, feeding the anode compartment of the fuel cell. Electromotive force is generated in the process of indirect oxidation and the generated electric energy can be used for maintenance of the wastewater treatment facility.

The MFCs were considered to be used for treating wastewater early in 1991 [45]. The power generated by MFCs in the wastewater treatment process can potentially halve the electricity needed in a conventional treatment process for aerating activated sludge. Later it was proposed to treat the wastewater in anaerobic digester to produce biogas and to yield volatile fatty acids, e.g. acetic, propionic and butyric. These acids feed the MFC to yield CO_2 and H_2O [46, 47].

Wastewater from breweries has been a favorite among researchers as a substrate in MFCs, primarily. The range of 3000–5000 mg of COD/dm^3 is typical for wastewater from breweries and it is approximately 10 times more concentrated than in domestic wastewater [48]. It could also be an ideal substrate for MFCs due to its high carbohydrate content and low ammonium nitrogen concentration [48]. Starch processing wastewater contains a relatively high content of carbohydrates, which can be potentially converted to useful products [43, 49].

The most spread application of microbial fuel cells is for treatment of domestic wastewater. It is comprehensively considered by Muñoz-Cupa et al. [50]. Considerable attention was paid to the treatment of landfill leachates resulting from the trickling rainwater through the waste layers in a landfill. For example, 242.3 Mt. of municipal solid waste was produced in the European Union in 2015, 62 Mt. of which were discarded in landfills [51]. Landfill leachate usually contain dissolved organic matter, inorganic macrocomponents, heavy metals, and xenobiotic compounds. These wastewater are associated with high organic load of up to 152000 mg COD/ dm^3 [52]. Such efforts have been made by Kjeldsen et al. [52], Galvez et al. [53] and Damiano [14]. It was reported that COD removal efficiency can reach more than 80%, with low power density (up to 125 W m^{-3}), cf. [50]. The energy output related to the destroyed COD is less than 0.04 kWh/kg COD. The Coulombic efficiency according to Faraday's law is very low, i.e. less than 15% [51]. However, the energy consumption for biodegradation of 1 kg COD in a traditional wastewater treatment facility is around 0.11 kWh/kg COD. It means that about 40% of the required energy can be supplied to the very process by MFC. The next benefit of MFC is the lower energy consumption for the substrate feed compared to the traditional methods where aeration and mixing are necessary.

There is a paper, reporting relatively high power densities (up to 10 W m^{-3}) in MFC after treatment of food leachate by gradual increase of pH in the feed [54]. At certain pH-values within the range of 6.3 and 7.6 the Coulombic efficiency reached 63%.

A combination of photochemical bioreactor and a fuel cell assembly for dairy waste removal was reported by Bolognesi et al. [55]. The Coulombic efficiencies were up to 21% with power density of 2.8 W m^{-3}.

The reported Coulombic efficiencies for MFC vary within a wide range (from 1 to 85%), depending on different conditions studied. The best current efficiencies found in the literature are for MFC operating with 0.4% sucrose solutions (between 71 and 85%) [56].

As can be seen, the power output and the current efficiency derived at MFC operation for wastewater treatment vary considerably depending on the operating conditions: type of substrate, microbes used (single cultures or consortia), cell immobilization, fuel cell construction, etc.

However, it is important to note, that the energy, consumed for MFC operation for this purpose is much less than the one for traditional aerated equipment, irrespective of the low Coulombic efficiency. Hence, the benefits of MFC use in these cases are multiple: first, it is the wastewater treatment and second, energy saving, avoiding the expensive mixing and aeration, with the use of simpler and cheaper equipment.

2.2 Microbial fuel cells with sulfate reduction

Another opportunity to use the MFC technique is to reduce sulfate ions in mine wastewater. They are usually acidic due to high sulfate content and they contain different heavy metals (copper, zinc, iron). Although they are in huge amounts their concentrations are low which make their chemical removal unfeasible from energy point of view. The large flow rates of the mine waters are an additional difficulty because large facilities are needed for water treatment with large energy demand. It is interesting to use the fuel cell technique to reduce sulfate as oxidizer instead of oxygen thus reducing them to sulfide [20–22]. Then the cations of heavy metals will be easily deposited as insoluble sulfides [27, 28, 57–59]. There is double effect – removal of heavy metals and sulfate and adjustment of pH for the drained water..The most studied sulfate-reducing bacteria are from the genera *Desulfovibrio, Desulfobacter, Desulfococcus,* etc. [28, 39, 40, 57].

The increase in electricity production in MFCs requires to optimize the inhibition of the metabolic pathways, which compete with electricity production [25].

The presence of sulfate and sulfide ions in anaerobic reactors hampers the growth of methanogenic archaea and justifies the use of sulfate and sulfate-reducing bacteria in the anodic space of MFC. It is important that the sulfate-reducing bacteria are able to transfer directly electrons to solid electrodes, thus reducing the competitive bulk redox processes in the anodic space. This technology associates the removal of both sulfate and the chemical oxygen demand (COD) with the production of electricity [22].

2.3 Microbial fuel cells with nitrate reduction

Nitrogen containing ions such as nitrate and nitrite occur widely in different process streams, coming from the extensive use of fertilizers in agriculture or production of explosives. The main problem is the reduction of nitrate to nitrite, both being toxic because of formation of methemoglobin [59] and carcinogenic N-nitrosamines [60] from nitrite.

The MFC with nitrate reduction differ, as all MFC, by the process involved on the other electrode, the bacterial culture used, and the construction of the cell. In most cases wastewater characterized by COD and nitrate polluted fluxes are treated. Many research efforts are directed to simultaneous nitrification and denitrification processes [23–25].

Systems working with a combination of sulfide and nitrate contaminated fluxes in the anode and cathode space of the fuel cell respectively are interesting for research [61–64] because of the simultaneous removal of two pollutants. Here there can be also nitrification and denitrification processes or only denitrification.

The following biochemical pathway of consecutive steps of nitrate-to-nitrite conversion and nitrite reduction to gaseous nitrogen found [65]:

$$NO_3^- \rightarrow NO_2^- \rightarrow NO(g) \rightarrow N_2O(g) \rightarrow N_2(g) \quad (3)$$

Denitrifying microorganisms are generally facultative anaerobes that could utilize chemically bound oxygen under anoxic conditions, e.g. in the nitrate ion, serving as electron acceptor instead of free oxygen.

It was pointed out by Wang et al. [65] that the microbial communities for denitrification are *Comamonadaceae* phylotypes and that the nitrification process could be explained by the high predominance of ammonia-oxidizing bacteria as *Nitrosomonas.* The existence of other potential heterotrophic denitrifiers as *Burkholderiadaceae, Alcaligenaceae* is also confirmed.

Pseudomonas denitrificans was used in [63, 65].

Pseudomonas sp. C27 is applied a pure culture, an autotrophic denitrifier, using sulfide as the only electron donor in a two-chambered MFC [64].

A technology that is particularly useful for nitrogen removal from wastewaters with low COD/N ratios is proposed in [23] to achieve simultaneous carbon and nitrogen removal from a single wastewater stream. For this purpose, the effluent from an acetate supplied MFC anode is directed to an aerobic stage for ammonium oxidation to nitrate. This stream is subsequently fed to the cathode of the MFC for denitrification. The removal rates are up to 2 kg COD m^{-3} day^{-1} and 0.41 kg NO_3-N m^{-3} day^{-1} with maximum power output of 34.6 W m^{-3} and a maximum current of 133 A m^{-3}.

In [25] a single chamber microbial fuel cell with a rotating biocathode was developed to remove simultaneously chemical oxygen demand (COD) and nitrogen with electric current generation. Carbon felt, on which exoelectrogenic bacteria grew, was used as the anode on the bottom and disks of the same material are the rotating cathode on the top. Under continuous regime with a feeding COD/N ratio of 5:1 the removal efficiencies of total organic carbon (TOC) and total nitrogen (TN) were 85.7 ± 7.4% and 91.5 ± 7.2%, respectively, and maximum power output of 585 mW m^{-3}.

In [63] a MFC that consists of two concentric cylindric compartments with effective volume of 300 ml each was investigated (**Figure 3**). The separating membrane was placed on the bottom of the inner cylinder. Activated and pyrolized paddling was used as anode. Granulated activated carbon was added in the cathode compartment in order to increase the electrode surface. The activated carbon was also used for immobilization of *Pseudomonas denitrificans (NBIMCC 1625)*. The results for the power generation in this construction are shown in **Figure 4**. The results are better for lower concentrations of sulfide and nitrate because of the inhibition on the bacterial activity at higher sulfate and nitrite concentrations.

The wetland fuel cells are an interesting solution for low cost wastewater treatment. Ge et al. [66] reported for simultaneous nitrate and phosphorus removal in a pyrite-based wetland-microbial fuel cell.

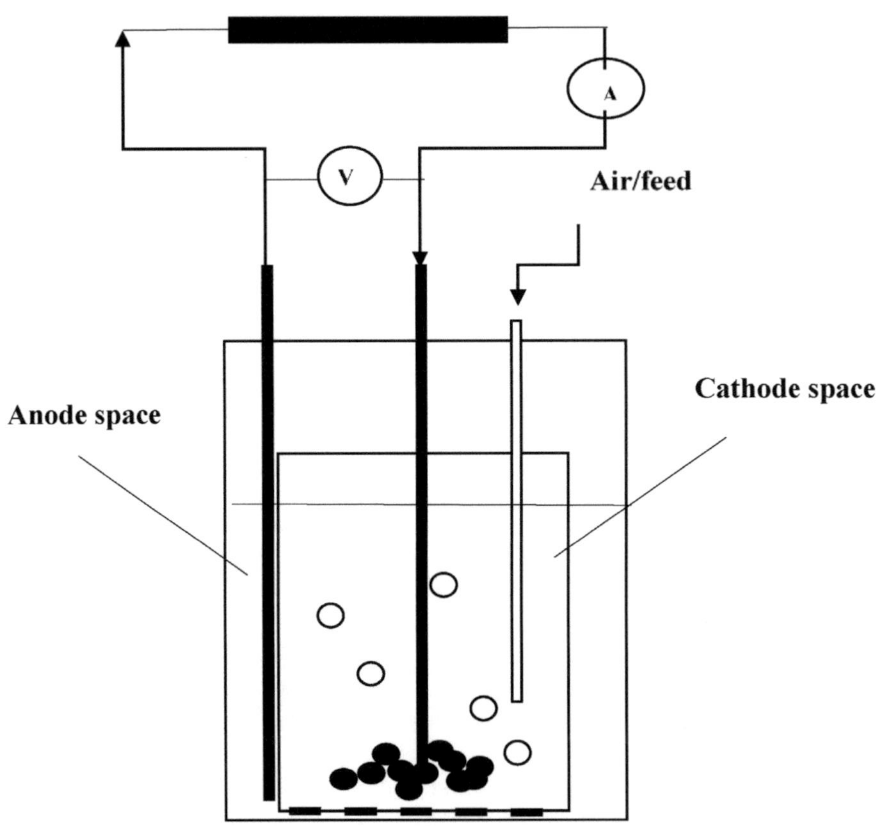

Figure 3.
Drawing of microbial fuel cell with cylindrical assembly of anode and cathode compartments.

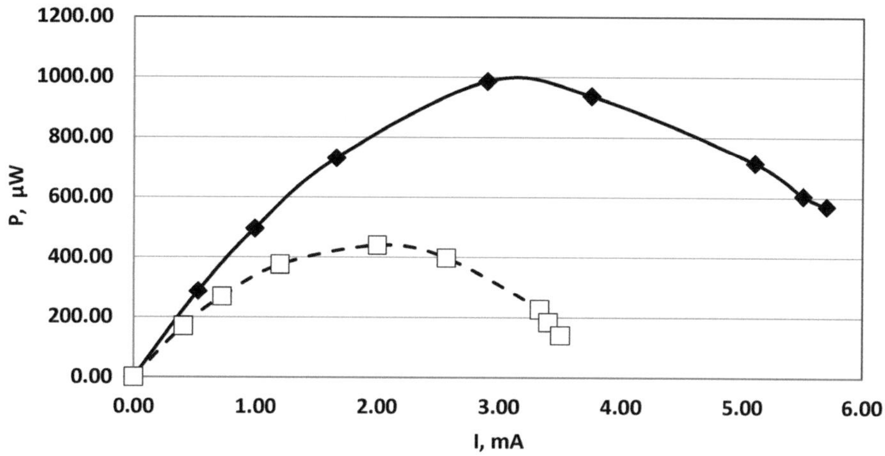

Figure 4.
Profile for power vs. current in cylindrical microbial fuel cell operating with sulfide and nitrate anions. (♦) – 3.5 mM sulfide, 3.5 mM nitrate; (□) – 15.6 mM sulfide, 8 mM nitrate.

3. Microbial electrolysis processes

There is a further development of the concept of microbial fuel cells. The aim is to treat wastewater by electrolysis facilitated by microbial process in the anodic space of electrolyzer with simultaneous production of hydrogen on the cathode [67–71].

Bioelectrochemical systems (BESs) can be regarded as electrochemical systems, in which at least one of the electrode reactions involves electrochemical interactions with electroactive bacteria. Most frequently it is the anodic reaction that requires the presence of certain microorganisms, yielding electrons from a biodegradable substrate to the anode. Some examples for such processes will be described briefly here.

3.1 Hydrogen production in microbial electrolysis cells

In contrast to the electrolytic production of hydrogen MEC require a small additional input of external electrical energy to facilitate hydrogen formation on the cathode [68–70]. The principle of MEC is similar to the traditional electrolysis ones but microbes grow on organic substrate in the anode compartment. Due to vital activity of the bacteria, chemical energy from organic matter in the wastewater is converted into electrical energy. The microorganisms used in MFCs are also applicable to MEC systems due to their similar anodic process [69].

There are papers, claiming to destroy landfill leachate and to produce hydrogen at the same time [71] and the COD removal efficiency is reported to reach 73%, whereas the hydrogen yield was up to 95%.

The voltage produced by the electroactive bacteria is not sufficient for water splitting and hydrogen release. That is why for microbial electrolysis a certain external voltage is applied. For the case of bacterially assisted microbial electrolysis cell (MEC) the hydrogen production increases with the applied voltage (i.e., 0.6–1.0 V) [72], which is much less than the theoretical voltage for abiotic water electrolysis (1.229 V).

3.2 Wastewater treatment by MEC

There are some studies on biological treatment of wastewater, containing recalcitrant organic pollutants, facilitated by electrolysis. There are also studies on the effect of Fenton pre-treatment on the performance of MEC in landfill leachate treatment [73] and the effect of co-substrate [52].

There is also an ambitious project to utilize the energy generated in MFC to treat wastewater in microbial electrolysis [74]. In [74] microbial consortium in activated sludge is used and it is coupled in parallel with MFCs for landfill leachate treatment for power generation. In this paper a 38% efficiency of COD and 90% of ammonia removal are attained. Although the current and power densities are low, this approach might save energy for wastewater treatment. The low voltage efficiencies can be overcome by coupling the MFCs in parallel or consecutively.

3.3 Microbial processes stimulated by constant electric fields

These processes are similar to the concept of MEC, but with the opposite aim. Whereas in the case of hydrogen production microbes facilitate electrolysis making it feasible in terms of energy, here the electric energy facilitates the natural redox processes in fermentation, thus reducing the energy consumption compared to aeration and mixing. Additional advantage could be the electrochemical removal of intermediates inhibiting the overall fermentation process. Some examples of this approach are described below.

3.3.1 Sorbitol-to-sorbose biotransformation

This biotransformation is a step in the traditional technology for L-ascorbic acid production (vitamin C) accomplished with the strain *Gluconobacter oxydans*. It is an aerobic process requiring high aeration rate and energy consumption. There was an effort to minimize energy consumption, applying constant electric field in the fermentation broth [75]. The results were positive: the constant electric field could replace partially intensive aeration and mixing, cf. **Figure 5**. The apparent volumetric mass transfer coefficient k_La was estimated as 6.5 h^{-1}, whereas the attained value at aeration and mixing is about 33 h^{-1}. It is visible from **Figure 5** that the most

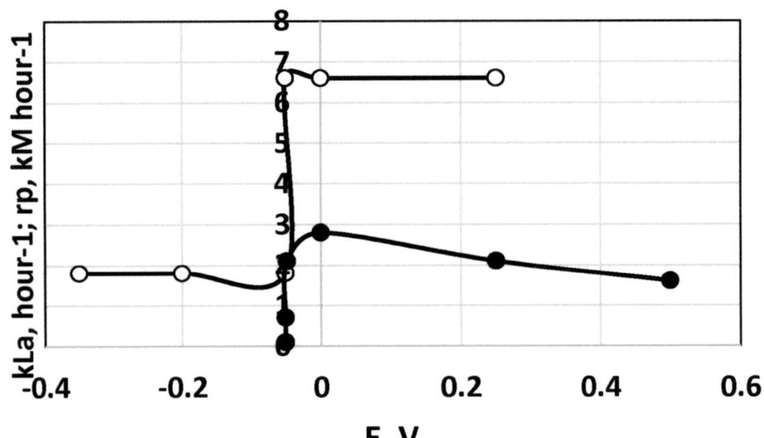

Figure 5.
Dependence of the production rate r_p (•) and the apparent volumetric mass transfer coefficient k_La (o). Initial sorbitol concentration 100 g dm^{-3}.

strongly manisfested effect was around a certain value of anode potential (i.e. – 0.05 V/S.H.E.), following the well-known "windows theory" according to which physical effects (like electric potential, current, temperature, etc.) have maximum impact on biological processes within a certain value range, like in an open window [76]. Calculations of Coulombic efficiency compared to the results of chemical analyses showed that the observed effect cannot be explained by purely electrochemical processes. The current efficiencies calculated by Faraday law were three orders of magnitude less than the expected ones by the chemical analyses. It was explained that the constant electric field affected the active sites of the enzyme sorbitol dehydrogenase. The optimum potential coincides with the standard potential of the co-enzyme couple $NAD^+/NADH^+$ (nicotinamide adenine dinucleotide), allowing speculation about a co-enzyme regeneration. However, this explanation is not convincible because of the very low measured Coulombic efficiency. Similar "window" effect was observed for the same biotransformation but in galvanostatic mode [77].

3.3.2 Biodegradation of 1,2-dichloroethane

This xenobiotic is frequently used in present practice as solvent, precursor for vinyl chloride and polyvinyl chloride production. Its biodegradation is usually accomplished by *Xanthobacter* strains, according to the following oxidative pathway [78]:

$$\underset{\text{1, 2-dichloroethane}}{CH_2Cl - CH_2Cl} \rightarrow \underset{\text{1-chloroethanol}}{H_2CCl - CH_2OH} \rightarrow \underset{\text{chloroactealdehyde}}{H_2CCl - CHO} \rightarrow \underset{\text{chloroacetate}}{H_2CCl - COO^-} \rightarrow \underset{\text{glycolate}}{CH_2OH - COO^-} \quad (4)$$

The ability of microbes to degrade this pollutant is limited at higher concentrations because of substrate inhibition and the accumulation of inhibitory intermediates, i.e. 1-chloroethanol, 1-chloroacetaldehyde and monochloroacetic acid. The idea was to enhance the intermediate reactions of oxidation and hence to remove the possible inhibitors as much as possible. Therefore, constant electric field for anodic oxidation was applied [79].

Experimental results on chloride release with and without application of electric field are shown in **Figure 6**. The results for the reference experiment with no electric field showed, that at higher substrate concentrations only one chlorine atom was released i.e. complete dechlorination was not attained. This fact corresponds to

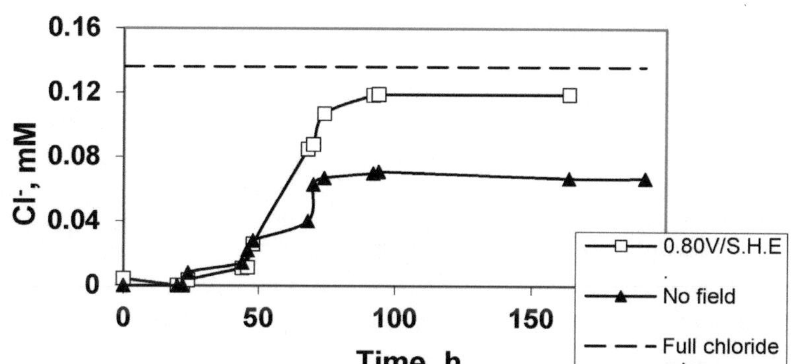

Figure 6.
Time profiles for chloride release for reference experiment and at constant electric field. Initial 1,2-DCE concentration 0.136 M.

the detected accumulation of 1-chloroethanol in the broth. When constant electric field was applied the chloride release was practically complete. In this case 1-chloroethanol was not detected in the broth. The optimum anode potential was 0.80 V/S.H.E. Then, a practically complete biodegradation took place even at high initial 1,2-DCE concentrations, up to 0.14 g dm^{-3}. At higher concentrations, e.g. 0.2 g dm^{-3}, only one chlorine atom was released, even when electric current was applied, possibly due to the dominating inhibition effects.

It was also observed that the effect of the electric field was not associated only with electrochemical process because of the very low Coulombic efficiencies, calculated from the electric current, being microampere-hours of the order of magnitude [80].

3.3.3 Phenol biodegradation

Phenol and its derivatives are considered as some of the most dangerous organic pollutants released into the environment due to human activity. Various methods have been applied to reduce its concentration in the waste streams to acceptable and harmless levels. In microbial degradation of phenol under aerobic conditions, the process is initiated by oxygenation, in which the aromatic ring is initially monohydroxylated at an *ortho*-position to the pre-existing hydroxyl group to form catechol by a monooxygenase phenol hydroxylase. Further, catechol is oxidized with *o*-benzoquinone formation [80] or to a cleavage of the benzene ring.

Depending on the type of strain, the catechol can undergo a ring cleavage that can occur either at *ortho*-position or at *meta*-position thus initiating the *meta*-pathway that leads to the formation of pyruvate and acetaldehyde, as suggested by Sridevi et al. [81].

The effects of constant electric fields on phenol biodegradation were studied by Ailijiang et al. [82], Dehghani et al. [83], Zhou et al. [84] and Beschkov et al. [85]. It was found in [85] that there is an optimum anode potential at which the specific bacterial growth rate for the strain *Pseudomonas putida* and its biodegrading ability are higher at anode potential of 0.8 V/S.H.E. corresponding to the standard redox potential for phenol oxidation [85]. The tests on enzyme activities for phenol hydrolase and 1,2-catechol dioxygenase showed that they are the best at the same potential. These data showed that only the *ortho-*pathway of benzene ring cleavage takes place. Still it appears that the effect of constant electric field is tends to be due to an effect on the enzyme activities rather than to electrochemical oxidation on the anode.

3.3.4 Bacterial denitrification

The essence of electrolytic stimulation of microbial degradation of nitrate consists in the microbial process in an electric field. The assumption was to facilitate the microbial activity by cathode production of hydrogen, which is a strong nitrate reducer [86–88]. Besides the galvanostatic mode of operation, there is a possibility to modify the bacterial activity by electrochemical stimulation under potentiostatic conditions [88]. Some results are shown in **Figure** 7. The advantage of the bioelectric stimulation is obvious. The optimum cathode potential (0.01 V/S.H.E.) coincides with the standard redox potential of the reaction:

$$NO_3^- + H_2O + 2e^- \rightarrow NO_2^- + 2OH^- \quad (5)$$

The reference experimental results showed that there was practically no electrochemical abiotic reduction of nitrate. However, again the Coulombic efficiency

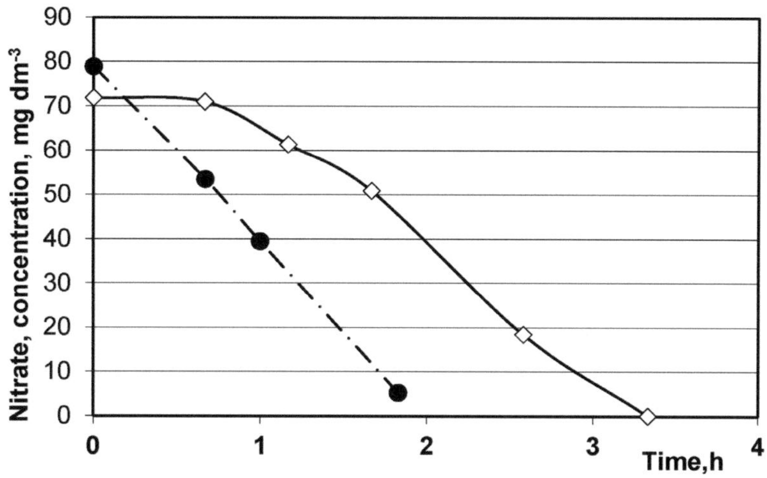

Figure 7.
Comparison of microbial denitrification by Pseudomonas denitrificans *at cathode potential of 0.232 V/S.H.E. with control experiment. (◇) – Control experiment; (•) – Process at constant electric field. Initial nitrate concentration, 80 mg dm*$^{-3}$*.*

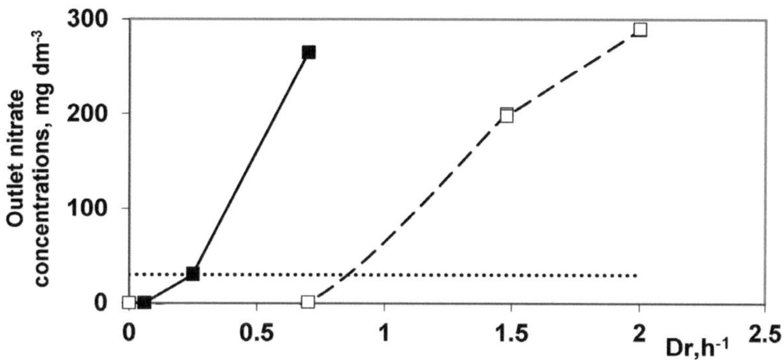

Figure 8.
Comparison of the performance of CSTR for bacterial denitrification by immobilized cells of Pseudomonas denitrificans *with and without constant electric field. Cathode potential 0.01 V/S.H.E. initial nitrate concentration – 300 mg dm*$^{-3}$*. Solid line, (■) – no electric field; dashed line (□) – With electric field, E = – 0.08 V/S.H.E. The horizontal dotted line denotes the dilution rates of 90% nitrate reduction.*

of this denitrification process is much lesser than the one corresponding to the chemical analyses. The biochemical stimulation effect of the electric field was confirmed by Field et al. [89] by electrochemical impedance spectroscopy, showing that the effect of enhanced microbial denitrification is the reduced activation energy of the reaction (5), i.e. due to changes in the active site of the enzyme nitrate reductase.

The strong denitrification effect of the constant electric field was confirmed for bacteria of the strain *Pseudomonas denitrificans* immobilized on a support of copolymer of acrylonitrile and acrylamide [90]. The effect of constant electric field is evident in continuous stirred tank reactor with immobilized cells, **Figure 8**. The break through dilution rate Dr = Q/V is more than twice higher than for control experiment without electric field.

One of the most valuable result of these studies is the fact that the bacteria can reduce nitrate to molecular nitrogen at much higher initial concentrations (up to 300 mg dm^{-3}) when electric field is applied, whereas without electric field the concentration upper limit for successful process is below 100 mg dm^{-3}. The strongest effect on nitrate and nitrite reduction is observed at cathode potential range from −0.1 to 0.1 V/S.H.E. Another important feature is the much higher flow rates of substrate feeds attainable when electric field is applied.

It was confirmed by experiments with cell free extract of *Pseudomonas denitrificans* after ultrasonic cell disintegration, that both nitrate reductase and nitrite reductase are stimulated significantly by the constant electric field at the similar cathode potential, namely 0.01 V/S.H.E., cf. [91].

There are recent studies to indicate the scientific interest toward this bioelectrochemical process [92, 93].

3.3.5 Electrostimulation of carbon dioxide conversion to methane and chemicals

There are efforts to recycle carbon dioxide into methane by MEC process [94–96]. First, carbon dioxide is stripped by alkaline absorption, by sodium hydroxide aqueous solution and afterwards passed through a MEC with methanogenic microbes in the cathode compartment. There bicarbonate ions are converted there into methane by some of the following reactions [97, 98]:

$$HCO_3^- + 9H^+ + 8e^- \rightarrow CH_4 + 3H_2O \quad (6)$$

$$HCO_3^- + 4H_2 + H^+ \rightarrow CH_4 + 3H_2O \quad (7)$$

$$CH_3COO^- + H_2O \rightarrow CH_4 + HCO_3^- \quad (8)$$

Various compounds can be used as electron donors, thus allowing to couple carbon dioxide recycling with additional wastewater treatment. Another new approach is to enhance methane production at anaerobic digestion by bioelectrochemical stimulation [94, 98, 99]. The effect consists in the easy removal or degradation of the excessive amounts of volatile fatty acids, e.g. propionic and butyric acids and to attain stable methanogenic process [94].

There are also efforts to reduce bioelectrochemically carbon dioxide to simple organic compounds [100, 101]. However, all these methods require energy input, loading the atmosphere with new amounts of carbon dioxide. An elegant solution will be to recycle carbon dioxide in new fuel cells using CO_2 as oxidizer and aqueous solutions of organic pollutants as a fuel. Hence, the benefit will be triple: removing carbon emissions from air, production of value-added organic compounds, thus saving fossil raw materials and cleaning wastewater at the same time.

All these achievements, as well as many more show serious challenges for further development of bioelectrochemistry in the biorefinery concept and contribute to the remedy the effects of the carbon emissions on climate change.

4. Conclusions

After reviewing the available literature on microbial fuel cells and microbial electrolysis the following conclusions can be drawn.

There are attractive challenges to use microbes to assist electric power generation during wastewater treatment by combining water purification with energy production. Although the power densities are rather low, the process is promising because of the lower energy demand compared to the traditional wastewater treatment technologies. Moreover, the generated energy can partially supply the equipment with energy. The purification rate of wastewater is promising even at low power production and Coulombic efficiency because of the lower energy demand compared to the traditional aerated process.

The microbial activities accompanied by redox processes can be used for energy saving in some electrolytic processes in the so-called microbial electrolysis cells. For example, hydrogen production by water splitting can be assisted by microbial processes with higher efficiency and lower energy consumption compared to traditional electrolysis.

The combination of fermentation redox processes with constant electric field can facilitate various microbial redox processes by stimulating enzyme catalysis under specific conditions, like fixed values of the electrode potential.

Bioelectrochemical effects can make it possible to recycle carbon dioxide to fuels and value-added chemicals.

All these applications will have practical use after increasing the process efficiencies and productivity. Higher power generation for MFC and lower energy consumption for MEC compared to the traditional processes must be attained. For this purpose, different approaches must be tested. On the one hand, suitable electrogenic microbes must be sought and applied for each particular case. On the other hand, highly active catalysts have to be invented and new designs of fuel cells, microbial electrolysis cells and appropriate electrodes must be developed.

Acknowledgements

The work on this article was supported by the project DN 17/4/2017 by the Fund for Scientific Research and by grant D01-214/2019, program Eplus, by the Ministry of Education and Science, Republic of Bulgaria.

List of abbreviations

BES	bioelectrochemical system
COD	chemical oxygen demand
DCE	dichloroethane
MEC	microbial electrolysis cell
MFC	microbial fuel cell
NAD^+	nicotinamide adenine dinucleotide
S.H.E.	standard hydrogen electrode

Author details

Venko Beschkov* and Elena Razkazova-Velkova
Institute of Chemical Engineering, Bulgarian Academy of Sciences, Sofia, Bulgaria

*Address all correspondence to: vbeschkov@yahoo.com

References

[1] Angenent L, Wrenn BA. Optimizing mixed-culture bioprocessing to convert wastes into bioenergy. In: Bioenergy. Herndon, VA. ASM Press; 2008. pp. 179-194

[2] Wu Q, Jiao S, Ma M, Peng S. Microbial fuel cell system: a promising technology for pollutant removal and environmental remediation. Environme ntal Science and Pollution Research. 2020;**27**:6749-6764

[3] Jabeen G, Farooq R. Microbial fuel cells and their applications for cost effective water pollution remediation. Proc. Natl. Acad. Sci., India, Sect. B Biol. Sci. 2017; **87:** 625–635 doi.org/10.1007/s40011-015-0683-x|.

[4] Logan BE, Rabaey K. Conversion of wastes into bioelectricity and chemicals by using microbial electrochemical technologies. Science. 2012;**337**(**6095**): 686-690

[5] O'Hayre RP, Cha SW, Colella WG, Prinz FB. Fuel Cell Fundamentals, 2nd ed.; J. Wiley & Sons, Inc.: N. Jersey, USA, 2009, p. 199. ISBN.978-0-470-25843-9.

[6] Lovley DR. Bug juice:harvesting electricity with microorganisms. Nature Reviews. 2006;**4**:497-508

[7] Logan BE. Microbial Fuel Cells. In: New Jersey.John Wiley & Sons,Inc. 2008

[8] Logan BE, Aelterman P, Hamelers B, Rozendal R, Schroede U, Keller J, et al. Microbial fuel cells: Methodology and technology. Environ. Sci. Technol. 2006;**40**:5181-5192

[9] Rabaey K, Boon N, Hofte M, Verstraete W. Microbial phenazine production enhances electron transfer in biofuel cells. Environ Sci Technol. 2005; **39**:3401-3408

[10] Hubenova Y, Hubenova E, Bakalska R, Mitov M. Redox interactions between dye 4-(E)-1-ethyl-4-(2-(4-hydroxyhaphtalen-1-yl)vinyl) quinolinium bromide and NAD^+/NADH. Bulg. Chem. Commun. 2018; 50: Special Issue D, 141-146.

[11] Hubenova Y, Hubenova E, Slavcheva E, Mitov M. The glyoxylate pathway contributes to enhanced extracellular electron transfer in yeast-based fuel cell. Bioelectrochem. 2017; **116**:10-16

[12] Hubenova Y, Bakalska R, Hubenova E, Mitov M. Mechanisms of electron transfer between styrylquinolinium dye and yeast in biofuel cell. Bioelectrochem. 2016;**112**: 158-165

[13] Kondaveeti SK, Seelam JS, Mohanakrishna G. Anodic Electron Transfer Mechanism in Bioelectrochemical Systems. In: Das D, editor. Microbial fuel cell: a bioelectrochemical system that converts waste to watts, Springer International Publishing (Cham, Switzerland) with Capital Publishung Company (New Delhi). 2018. pp. 87-100

[14] Damiano L. Electricity production from the management of municipal solid waste leachate with microbial fuel cells. Master's Theses and Capstones. 2009;**443** https://scholars.unh.edu/thesis/443

[15] Rabaey K, Boon N, Siciliand SD, Verhaege M, Verstraete W. Biofuel cells select for microbial consortia that self-mediate electron transfer. Applied and Environmental Microbiology. 2004;**70**: 5373-5382

[16] Reguera G, Nevin KP, Nicoll JS, Covalla SF, Woodard TL, Lovley DR. Biofilm and nanowire production leads

to increased current in Geobacter sulfurreducens fuel cells. Applied and Environmental Microbiology. 2006;**72**: 7345-7348

[17] Pant D, Van Bogaert G, Diels L, Vanbroekhoven K. A review of the substrates used in microbial fuel cells (MFCs) for sustainable energy production. Bioresource Technology. 2010;**101**:1533-1543

[18] Oluwaseun A. Bioremediation of petroleum hydrocarbons using microbial fuel cells, PhD Thesis, University of Westminster. Faculty of Science and Technology. 2015:96-132

[19] Oluwaseun A, Keshavarz T, Kyazze G. Enhanced biodegradation of phenanthrene using different inoculum types in a microbial fuel cell. Eng. Life Sci. 2014;**14**:218-228

[20] Lee D, Lee C, Chang J, Liao Q, Su A. Treatment of sulfate/sulfide-containing wastewaters using a microbial fuel cell: single and two-anode systems. International Journal of Green Energy. 2015;**12**:998-1004. DOI: 10.1080/15435075.2014.910780

[21] Lee D, Lee C, Chang J. Treatment and electricity harvesting from sulfate/sulfide-containing wastewaters using microbial fuel cell with enriched sulfate-reducing mixed culture. Journal of Hazardous Materials. 2012;**243**:67-72

[22] Lee D, Liu X, Weng H. Sulfate and organic carbon removal by microbial fuel cell with sulfate-reducing bacteria and sulfide-oxidising bacteria anodic biofilm. Bioresource Technology. 2014; **156**:14-19

[23] Virdis B, Rabaey K, Yuan Z, Keller J. Microbial fuel cells for simultaneous carbon and nitrogen removal. Water Research. 2008;**42**:3013-3024

[24] Virdis B, Rabaey K, Rozendal A, Yuan Z, Keller J. Simultaneous nitrification, denitrification and carbon removal in microbial fuel cells. Water Research. 2010;**44**:2970-2980

[25] Zhang G, Zhang H, Zhang C, Zhang G, Yang F, Yuan G, et al. Simultaneous nitrogen and carbon removal in a single chamber microbial fuel cell with a rotating biocathode. Process Biochemistry. 2013;**48**:893-900

[26] Sotres A, Cerrillo M, Viñas M, Bonmatí A. Nitrogen removal in a two-chambered microbial fuel cell: Establishment of a nitrifying–denitrifying microbial community on an intermittent aerated cathode. Chemical Engineering Journal. 2016;**284**:905-916

[27] Rodrigues ICB, Leão VA. Producing electrical energy in microbial fuel cells based on sulphate reduction: a review. Environmental Science and Pollution Re search. 2020;**27**:36075-36084

[28] Bratkova S, Lavrova S, Angelov A, Nikolova K, Ivanov R, Koumanova B. Treatment of wastewaters containing Fe, Cu, Zn and As by microbial hydrogen sulfide and subsequent removal of COD. N and P. Journal of Chemical Technology and Metallurgy. 2018;**53**:245-257 https://dl.uctm.edu/journal/node/j2018-2/11_17_179_p_245_257.pdf

[29] Loukanov A, Angelov A, Takahashi Y, Nikolov I, Nakabayashi S. (2019), Carbon nanodots chelated with metal ions as efficient electrocatalysts for enhancing performance of microbial fuel cell based on sulfate reducing bacteria. Colloids and Surfaces A: Physicochemical and Engineering Aspects. 2019;**574**:52-61. DOI: 10.1016/j.colsurfa.2019.04.067

[30] Park DH, Zeikus JG. Impact of electrode composition on electricity generation in a single-compartment fuel cell suing *Shewanella putrefaciens*. Appl Microbiol Biotechnol. 2002;**59**:58-61

https://doi.org/10.1007/s00253-002-0972-1

[31] Park DH, Zeikus JG. Improved fuel cell and electrode designs for producing electricity from microbial degradation. Biotechnol Bioeng. 2003;**81**:348-355 https://doi.org/10.1002/bit.10501

[32] Liu H, Cheng S, Logan BE. Power generation in fed-batch microbial fuel cells as a function of ionic strength, temperature, and reactor configuration. Environ Sci Technol. 2005;**39**:5488-5493 https://doi.org/10.1021/es050316c

[33] Mitov M, Chorbadzhiyska E, Nalbandian L, Hubenova Y. Nickel-based electrodeposits as potential cathode catalysts for hydrogen production by microbial electrolysis. J. Power Sources. 2017;**356**:467-472

[34] Hubenova Y, Rashkov R, Buchvarov V, Babanova S, Mitov M. Nanomodified NiFe-and NiFeP-carbonfelt as anode electrocatalysts in yeast-biofuel cell. J Mater Sci. 2011;**46**: 7074-7081

[35] Logan B, Cheng S, Watson V, Estadt G. Graphite fiber brush anodes for increasing power production in air-cathode microbial fuel cells. Environmental Science and Technology. 2007;**41**:3341-3346

[36] Liu H, Logan BE. Electricity generation using an air-cathode single chamber microbial fuel cell in the presence and absence of a proton exchange membrane. Environmental Science and Technology. 2004;**38**: 4040-4046

[37] Liu H, Ramarayanan R, Logan BE. Production of electricity during wastewater treatment using a single chamber microbial fuel cell. Environmental Science and Technology. 2004;**38**:2281-2285

[38] ter Heijne A. Improving the cathode of a microbial fuel cell for efficient electricity production., PhD Thesis. Wageningen University 2010; pp. 131-134.

[39] Houari A, Ranchou-Peyruse M, Ranchou-Peyruse A, Dakdaki A, Guignard M, Idouhammou L, et al. *Desulfobulbus oligotrophicus* sp. nov., a sulfate-reducing and propionate-oxidizing bacterium isolated from a municipal anaerobic sewage sludge digester. Int J Syst Evol Microbiol. 2017; **67**:27-281

[40] Kang CS, Eaktasang N, Kwon DY, Kim HS. Enhanced current production by *Desulfovibrio desulfuricans* biofilm in a mediator-less microbial fuel cell. Bioresource Technology. 2014;**165**:27-30

[41] Rabaey K, Rodríguez J, Blackall LL, Keller J, Gross P, Batstone D, et al. Microbial ecology meets electrochemistry: electricity-driven and driving communities. ISME J. 2007;**1**:9-18

[42] Cházaro-Ruiz LF, López-Cázares MI, González I, Toriz Y, Alatriste-Mondragon F, Santana M, et al. Improving substrate consumption and decrease of growth yield in aerobic cultures of *Pseudomonas denitrificans* by applying low voltages in bioelectric systems. Appl Biochem Biotechnol. 2020;**190**:1333-1348 https://doi.org/10.1007/s12010-019-03168-x

[43] Niessen J, Schroder U, Scholz F. Exploiting complex carbohydrates for microbial electricity generation — a bacterial fuel cell operating on starch. Electrochem Commun. 2004;**6**:955-958

[44] Niessen J, Harnisch F, Rosenbaum M, Schroder U, Scholz F. Heat treated soil as convenient and versatile source of bacterial communities for microbial electricity generation. Electrochem Commun. 2006;**8**:869-873

[45] Habermann W, Pommer EH. Biological fuel cells with sulphide

storage capacity. Applied Microbiology and Biotechnology. 1991;**35**:128-133

[46] Gude VG. Integrating bioelectrochemical systems for sustainable wastewater treatment. Clean Technologies and Environmental Policy. 2018 https://doi.org/10.1007/s10098-018-1536-0

[47] Du Z, Li H, Gu T. A state of the art review on microbial fuel cells: A promising technology for wastewater treatment and bioenergy. Biotechnology Advances. 2007;**25**:464-482

[48] Vijayaraghavan K, Ahmad D, Lesa R. Electrolytic treatment of beer brewery wastewater. Ind. Eng. Chem. Res. 2006;**45**:6854-6859

[49] Jin B, van Leeuwen HJ, Patel B, Yu Q. Utilisation of starch processing wastewater for production of microbial biomass protein and fungal α-amylase by *Aspergillus oryzae*. Bioresource Technology. 1998;**66**: 201-206

[50] Muñoz-Cupa C, Hu Y. Xu C (Charles), Bassi A, An overview of microbial fuel cell usage in wastewater treatment, resource recovery and energy production. Scie nce of the Total Environment. 2021;**754**: 142429

[51] Bolognesi S, Cecconet D, Callegari A, Capodaglio AG. Bioelectrochemical treatment of municipal solid waste landfill mature leachate and dairy wastewater as co-substrates. Environ Sci Pollut Res. 2020 https://doi.org/10.1007/s11356-020-10167-7

[52] Kjeldsen P, Morton A, Barlaz A, Rooker P, Baun A, Ledin A, et al. Present and long-term composition of MSW landfill leachate: A Review. Critical Reviews in Environmental Science and Technology. 2002;**32**: 297-336

[53] Gálvez A, Greenman J, Ieropoulos I. Landfill leachate treatment with microbial fuel cells; scale-up through plurality. Bioresource Technology. 2009;**100**:5085-5091. DOI: 10.1016/j.biortech.2009.05.061

[54] Li XM, Cheng KY, Wong JW. Bioelectricity production from food waste leachate using microbial fuel cells: effect of NaCl and pH. Bioresource Technology. 2013;**149**:452-458. DOI: 10.1016/j.biortech.2013.09.037

[55] Bolognesi S, Cecconet D, Callegari A, Capodaglio AG. Combined microalgal photobioreactor/microbial fuel cell system: Performance analysis under different process conditions. Environ Res. 2020;**192**:110263. DOI: 10.1016/j.envres.2020.110263

[56] Devasahayam M, Masih SA. Microbial fuel cells demonstrate high coulombic efficiency applicable for water remediation. Indian J Exp Biol. 2012;**50**:430-438

[57] Bratkova S, Ivanov R, Gerginova M, Peneva N, Angelov A, Alexieva Z. Rhizosphere microflora of sediment plant microbial fuel cells, In: A. Méndez-Vilas, editor. Exploring microorganisms: recent advances in applied microbiology. Brown Walker Press, 2018. 40-43. https://www.universalpublishers.com/book.php?method=ISBN&book=1627346236.

[58] Cai J, Qaisar M, Sun Y, Wang K, Lou J, Wang R. Coupled substrate removal and electricity generation in microbial fuel cells simultaneously treating sulfide and nitrate at various influent sulfide to nitrate ratios. Bioresource Technology. 2020; 306: 123174.

[59] Fan AM, Steinberg VE. Health implications of nitrate and nitrite in drinking water: an update on methemoglobinemia occurrence and reproductive and developmental

toxicity. Regul Toxicol Pharmacol. 1996; **23**:35-43

[60] Forman D. Nitrate Exposure and Human Cancer. In: Bogárdi I, Kuzelka RD, Ennenga WG, editors. Nitrate Contamination. NATO ASI Series (Series G: Ecological Sciences). Vol. 30. Berlin, Heidelberg: Springer; 1991 https://doi.org/10.1007/978-3-642-76040-2_20

[61] Cai J, Zheng P. Simultaneous anaerobic sulfide and nitrate removal in microbial fuel cell. Bioresource Technology. 2013;**128**:760-764

[62] Stefanov S, Razkazova-Velkova E, Martinov M, Parvanova-Mancheva T, Beschkov V, Sulfide and nitrate driven fuel cell. Chemical and biochemical denitrification. Bulgarian Chemical Communications, 2018; 50: Special Issue B. 123 – 129.

[63] Jordanova M, Yankov D, Stefanov S, Razkazova-Velkova E. Microbial fuel cell for metal sulfide oxidation and nitrate reduction. Part I. Preliminary investigation of electrogenic properties. Bulg. Chem. Commun. 2020; 52: Special Issue A. 69-73.

[64] Sotres A, Cerrillo M, Viñas M, Bonmatí A. Nitrogen removal in a two-chambered microbial fuel cell: Establishment of a nitrifying–denitrifying microbial community on an intermittent aerated cathode. Chemical Engineering Journal. 2016; **284**:905-916

[65] Wang JH, Baltzis BC, Lewandowski GA. Fundamental denitrification kinetic studies with Pseudomonas denitrificans. Biotechnology Bioengineering. 1995;**47**: 26-41

[66] Ge X, Cao X, Song X, Wang Y, Si Z, Zhao Y, et al. Bioenergy generation and simultaneous nitrate and phosphorus removal in a pyrite-based constructed wetland-microbial fuel cell. Bioresource Technology. 2020;**296**:122350

[67] Shen RX, Zhao LX, Lu JW, Watson J, Si BC, Chen X, et al. Treatment of recalcitrant wastewater and hydrogen production via microbial electrolysis cells. Int J Agric & Biol Eng. 2019;**12**:179-189

[68] Azwar MY, Hussain MA, Abdul-Wahab AK. Development of biohydrogen production by photobiological, fermentation and electrochemical processes: A review. Renewable and Sustainable Energy Reviews. 2014;**31**:158-173

[69] Rozendal RA, Hubertus V, Hamelers M, Euverink GLW, Metz SJ, Buisman CJN. Principle and perspectives of hydrogen production through biocatalyzed electrolysis. Inte rnational Journal of Hydrogen Energy. 2006;**31**:1632-1640

[70] Escapa A, Mateos R, Martínez EJ, Blanes J. Microbial electrolysis cells: An emerging technology for wastewater treatment and energy recovery. From laboratory to pilot plant and beyond. Renewable and Sustainable Energy Reviews. 2016;**55**:942-956

[71] Hassan M, Sotres Fernandez A, San Martin I, Xie B, Moran A. Hydrogen evolution in microbial electrolysis cells treating landfill leachate: dynamics of anodic biofilm. International Journal of Hydrogen Energy. 2018;**43**:13051-13063

[72] Rozendal RA, Hamelers HVM, Molenkamp RJ, Buisman CJN. 2007. Performance of single chamber biocatalyzed electrolysis with different types of ion exchange membrances. Water Research. 2007;**41**:1984-1994

[73] Huang T, Liu L, Tao J, Zhou L, Zhang S. Microbial fuel cells coupling with the three-dimensional electro-Fenton technique enhances the degradation of methyl orange in the

wastewater. Environmental Science and Pollution Research. 2018;**25**: 17989-18000

[74] Feng Q, Xu L, Xu Y, Liu C, Lu Y, Wang H, et al. Treatment of aged landfill leachate by a self-sustained microbial fuel cell-microbial electrolysis cell system. Int. J. Electrochem. Sci. 2020;**15**:1022-1033. DOI: 10.20964/2020.01.19

[75] Mitov S, Beschkov V. The biochemical oxidation of D-sorbitol to L-sorbose at constant electric current. Bioelectrochemistry & Bioenergetics. 1992;**28**:435-442

[76] Berg H. Possibilities and problems of low-frequency weak electromagnetic field in cell biology. Bioelectrochemistry & Bioenergetics. 1995;**38**:153-159

[77] Beschkov VN, Peeva LG. Effect of electric current passing through the fermentation broth of a strain Acetobacter suboxydans. Bioelectrochemistry & Bioenergetics. 1994;**34**:185-188

[78] Janssen D B., Pries F, van der Ploeg J, Kazemier B, Terpstra P, Witholt B (1989) Cloning of 1, 2-dichlor-oethane degrading genes of *Xanthobacter autotrophicus* GJ10 and expression and sequencing of the*dhlA* gene. J. Bacteriol. 1989; **171:** 6791–6799.

[79] Vasileva E. Parvanova-Mancheva, Beschkov V, Influence of constant electric field on dehalogenation capacity of the strain *Xanthobacter autotrophicus* GJ 10. Journal of International Scientific Publications, Ecology & Safety. 2017;**11**: 91-100 www.scientific-publications.net

[80] Dey SK. Mukherjee A (2016) Catechol oxidase and phenoxazinone synthase: Biomimetic functional models and mechanistic studies. Coord Chem Rev. 2016;**310**:80-115 https://doi.org/10.1016/j.ccr.2015.11.002

[81] Sridevi V, Chandana Lakshmi MVV, Manasa M. Sravani M (2012) Metabolic pathways for the biodegradation of phenol. Int J Eng Sci Adv Technol. 2012; **2**:695-705

[82] Ailijiang N, Chang J, Liang P, Li P, Wu Q, Zhang X, Huang X. Electrical stimulation on biodegradation of phenol and responses of microbial communities in conductive carriers supported biofilms of the bioelectrochemical reactor. Bioresource Technology. 2016; **201:** 1–7. https://doi.org/10.1016/j.biort ech

[83] Dehghani S, Rezaee A, Moghiseh Z. Phenol biodegradation in an aerobic fixed-film process using conductive bioelectrodes: Biokinetic and kinetic studies. Desalination and Water Treatment. 2018;**105**:126-131

[84] Zhou L, Yan X, Yan Y, Li T, An J, Liao CM, et al. Electrode potential regulates phenol degradation pathways in oxygen-diffused microbial electrochemical system. Chem Eng J. 2019;**381**:122663 https://doi.org/10.1016/j.cej.2019.12266 3

[85] Beschkov V, Alexieva Z, Parvanova-Mancheva T, Vasileva E, Gerginova M, Peneva N, et al. Phenol biodegradation by the strain *Pseudomonas putida* affected by constant electric field. International Journal of Environmental Science and Technology. 2020;**17**:1929-1936. DOI: 10.1007/s13762-019-02591-1

[86] Kuroda M, Watanabe T, Umedu Y. (1997) Simultaneous COD removal and denitrification of wastewater by bio-electro reactors. Water. Sci Technol. 1997;**35**:161-168

[87] Feleke Z, Araki K, Sakakibara Y, Watanabe T, Kuroda M. Selective reduction of nitrate to nitrogen gas in a biofilm-electrode reactor. Water Research. 1998;**32**:2728-2734

[88] Beschkov V, Velizarov S, Agathos SN, Lukova V. Bacterial

denitrification of wastewater stimulated by constant electric field. Biochemical Engineering Journal. 2004;**17**:141-145

[89] Field SJ, Thornton NP, Anderson LJ, Gates AJ, Reilly A, Jepson BJN, Richardson DJ, George SJ, Cheesman MR, Butt JN. Reductive activation of nitrate reductases. Dalton Trans. 2005; **21:** 3580–3586. https://doi.org/10.1039/b5055 30j

[90] Parvanova-Mancheva T, Beschkov V. Microbial denitrification by immobilized bacteria *Pseudomonas denitrificans* stimulated by constant electric field. Biochemical Engineering Journal. 2009;**44**:208-213. DOI: 10.1016/j.bej.2008.12.005

[91] Parvanova-Mancheva T, Beschkov V, Sapundzhiev T. Modeling of biochemical nitrate reduction in constant electric field. Chemical and Biochemical Engineering Quarterly. 2009;**23**:67-75

[92] Cadena Ramírez A, Texier AC, González I, Gómez Hernández J. Implications of electric potentials applied on a denitrifying process. Environmental Technology. 2019;**40**: 2747-2755 https://doi.org/10.1080/09593330.2018.1450447

[93] Dehghani S, Rezaee A. Biological denitrification using microbial electrochemical technology: a perspective of materials, the arrangement of electrodes and energy consumption. Desalination and Water Treatment. 2020;**178**:155-162

[94] Liu D, Zeppilli M, Villano M, Buisman C, ter Heijne A. Methane production at biocathodes: principles and applications, In: Bioelectrosynthesis: principles and technologies for value-added products, First Edition. Edited by Aijie Wang, Wenzong Liu, Bo Zhang, and Weiwei Cai. 2020, Wiley-VCH Verlag GmbH & Co. KGaA. Chapter;**5**: 129-159

[95] Hara M, Onaka Y, Kobayashi H, Fu Q, Kawaguchi H, Vilcaez J, et al. Ene rgy Procedia. Mechanism of electromethanogenic reduction of CO_2 by a thermophilic methanogen. 2013;**37**: 7021-7028

[96] Villano M, Aulenta F, Ciucci C, Ferri T, Giuliano A, Majone M. Bioelectrochemical reduction of CO2 to CH4 via direct and indirect extracellular electron transfer by a hydrogenophilic methanogenic culture. Bioresource Technology. 2010;**101**:3085-3090

[97] van Eerten-Jansen MCAA, Jansen NC, Plugge CM, de Wilde V, Buisman CJN, ter Heijne A. Analysis of the mechanisms of bioelectrochemical methane production by mixed cultures. J. Chem. Technol. Biotechnol. 2015;**90**: 963-970

[98] Liu D, Zhang L, Chen S, Buisman CJN, ter Heijne A. Bioelectrochemical enhancement of methane production in low temperature anaerobic digestion at 10∘C. Water Research. 2016;**99**:281-287

[99] Tartakovsky B, Mehta P, Bourque JS, Guiot SR. Electrolysis-enhanced anaerobic digestion of wastewater. Bioresource Technology. 2011;**102**:5685-5691

[100] Su M, Jiang Y, Zhang Y, Gao P, Li D. Coupled bioelectrochemical system for reducing CO_2 to simple organic compounds in the presence of H_2. Chinese J. Appl. Environ. Biol. 2013; **19**:827-832

[101] Li H, Opgenorth PH, Wernick DG, Rogers S, Wu TY, Higashide W, et al. Integrated electromicrobial conversion of CO_2 to higher alcohols. Science. 2012; **335**:1596. DOI: https://doi.org/10.1126/science.1217643.

Chapter 7

Electrode Material as Anode for Improving the Electrochemical Performance of Microbial Fuel Cells

Asim Ali Yaqoob, Mohamad Nasir Mohamad Ibrahim and Khalid Umar

Abstract

The energy generation without causing environmental pollution is a unique idea to make a better survival for human beings. In this regard, microbial fuel cells (MFCs) have been considered to be eco-friendly and efficient technology to produce renewable energy. The operations and functioning of MFCs technology were affected by many factors but the electrodes are the most essential and significant aspects in MFCs. Moreover, a wide variety of electrodes and MFCs configurations have been developed to enhance the electrochemical performance of MFCs. The carbon materials (graphite, graphene etc.) were commonly used for the electrode fabrication, due to some unique properties such as high conductivity, good thermal stability, high surface area, good mechanical power etc. In this chapter, different electrode materials, used for anode fabrication were summarized to reveal the performance/efficiency toward the generation of electricity. Finally, the electrochemical characterizations tool, current challenges, and future perspectives of the electrode in MFCs were discussed briefly.

Keywords: Microbial fuel cells, Anode electrodes, Energy, Electrochemical characterizations, Anode oxidation

1. Introduction

Due to the increasing worldwide population, industries, etc. the commercial, and domestic demands of energy are on their rising trends. The rapidly growing population and industrialization have carried out a vast influence on environment contamination such as water, air pollution [1]. Therefore, it has been essential to overcome the energy demand without environmental pollution. In past decades, wastewater was considered as an efficient source for generation of energy. Hence, there is a crucial need to introduce a method by improving energy recovery from water resources without high expenditure. To respond to this inefficiency, the scientific community introduced a new method called microbial fuel cells (MFCs), considered as the most economical and environmentally stable process. Logan et al. [2] defined the MFCs as a most prospective approach to generate energy along with treatment of wastewater by using microbes as catalysts. Similarly, Nikhil et al. [3]

studied that the MFCs have the capability to generate electricity along with the treatment of wastewater and raised it to be an ideal solution for energy and water issues. Usually, MFCs presented extensive advantages as compared to other conventional techniques such as less activated sludge produced from wastewater, energy input is not present for aeration and the system works at various pH, temperature and substrate to generate electricity through using microorganism electro-catalyst. Despite all developments in MFCs, even an ideal improvement in power density was observed earlier but still practical application has not been implemented in the market due to the less electric efficiency. There are several factors such as electrodes (anode & cathode), incolumn source, proton exchange membrane, MFCs scale, design and organic substrate etc. which directly affect the operation of MFCs [4]. Moreover, the MFCs have two basic types of electrodes in the system such as anode and cathode system. In all factors, one of the most important aspects is the electrode material, particularly anode electrode and its working efficiency. The lower transportation rate of electrons becomes a major problem which decreases the electric performance of the MFCs. In early stage, a large variety of electrodes (activated carbon cloth, carbon felt, cloth, rod, mesh, fiber, paper, brushes, 3D graphite, glassy carbon, granular graphite, graphite block, graphite felt, reticulated vitreous carbon and graphite oxide) have been widely studied which were made using different materials [5]. Both types of electrodes play a significant role in the performance of the system but regarding in terms of functionality, anode attracted a lot of attention by researchers [6].

Furthermore, the organic substrate is oxidized by electrochemical active microbes usually known as exoelectrogens which are present on the surface of anode. The exoelectrogens produce the electrons/protons and significantly reduce the toxic effect of pollutants from wastewater. The redox reaction is thermodynamically significant, releasing electrons by microbes' stream from anode toward the cathode electrode through the provided outside circuit, and thus producing electricity. Therefore, it is not difficult to say that the MFCs performance depends on efficiency of the electrode material (anode and cathode material) [7, 8]. Furthermore, to promote the rate of organic substrate oxidation at the surface of anode electrode, a significant development is required for anode material to increase the electron transportation. The high oxidation rate led to the high generation of electrons which can travel toward the cathode so an efficient medium (anode material) can effectively transfer the electrons from anode to cathode chamber. Recently, Yaqoob et al. [7] reported the modified graphene/polyaniline nanocomposite as anode vs. graphite as cathode in MFCs to enhance electron transportation. A.A.Y. et al. reported the 87.71 mA m^{-2} current density which was higher as compared to simple graphite anode. In another study, Yaqoob et al. [8] also proved that the modification of anode electrodes can bring high electron transfer which led to high performance of MFCs. A.A.Y. et al. also used the graphene/ZnO composite as anode and reported 142.98 mA m^{-2} current density. In the previous literature, ElMekawy et al. [9] studied the graphene-based electrodes (anode & cathode) and compared the performance in terms of electricity. The highest achieved power density was reported as 4 W m^{-2} by using graphene-based anode electrodes vs. carbon brushes as cathode. Similarly, modified cathode (graphene-based) was offered with a power density as 3.5 W m^{-2}. Therefore, this comparatively higher power density by graphene modified anode, was considered more efficient as compared to graphene-modified cathode. The reason for higher power density in case of graphene modified anode is due to the high electron transportation toward cathode. While in graphene-based cathode the energy was low due to poor rate of electron transportation from anode. Therefore, the generated electron required an effective anode material to transfer the electrons rapidly toward the cathode. Moreover, the authors also proposed that graphene anode modification is more important than the cathode electrode to get better performance. The anode material braked the progress of MFCs at practical application due to low conductive material, bacterial toxicity, corrosive and cost issues. The significant value, role of anode and above-mentioned issues urge the utilization of highly conductive material as anode material to overcome these problems. Generally, some properties such as high surface area, high conductivity, biocompatibility, cost effectiveness, high mechanical and thermal stability are essential required parameters for high performance of anode electrode [9]. To overcome these current challenges, the utilization of high-performance material will be considered as a great idea. Although the materials used mostly are carbon-based, less effort seems to develop graphene and its derivatives-based electrodes. Therefore, the present chapter describes the role and working mechanism of anode oxidation in MFCs. Presently, several efforts are still ongoing to improve the anode material as electrodes for high performance related to MFCs. Moreover, this chapter will be proved very useful for the researcher to develop an idea for further improvement regarding anode electrodes. However, the basic required properties and future recommendations for high performance anode are briefly summarized in this article.

2. Working mechanism of microbial fuel cells via anode oxidation

Many microbial species are well known exoelectrogens in MFCs and they can oxidize the organic substrate to produce the electron and proton. The *Aeromonas hydrophila, Shewanella spp. Geobacter spp., Clostridium butyricum Enterococcus gallinarum* and *Rhodoferax ferrireducens* are mostly reported exoelectrogens in MFCs studies [10, 11]. Some bacterial species such as *Geobacter species* also have the ability for electron generation properties [10–12]. For example, *Geobacter* based biofilms can work as transistors and supercapacitors due to the conductive phenomenon. The exoelectrogens can also pass and produce electrons, and this production of current is generated because of their respiration phenomenon. Usually, the anode is not part of the ordinary aquatic environment; so, microbes typically act as a transference medium for electrons to insoluble natural electron acceptors [13]. Throughout the reduction of pollutants, the pili of *Geobacter* are vigorously voiced. The microbes are ascribed to anode electrodes and produce the biofilms. The microbes are responsible to transfer the electrons toward the anode surface as compared to natural insoluble mediators as electron acceptors. In microbial metabolism and bioelectrogenesis, organic substrates comprising carbohydrates are generally used in MFCs. These carbohydrate-based organic substrate molecules generate the acetyl coenzyme A by entering into the glycolysis process and further respective ways. This acetyl coenzyme A enters the tricarboxylic acid cycle (Kreb's cycle). All these processes take place in the cytoplasm and one complete round of Krebs cycle generate three molecules of reduced nicotinamide adenine dinucleotide (NADH) and one molecule of reduced flavin adenine dinucleotide ($FADH_2$) with CO_2 as a by-product [14–16]. To generate the adenosine triphosphate (an energy carrier molecule), the NADH and $FADH_2$ act as electron transporters, and then they permit their electrons toward the electron transport chain (ETC). The ETC operates in the cell membrane (inner and outer membranes and periplasm) and these electrons move through consequent protein channels (NADH dehydrogenase, ubiquinone, coenzyme Q, and cytochromes) of the ETC and lastly captured by electron acceptors [17]. Later, they were transferred to the cathode electrode from the anode via an external circuit.

This entire cycle produces approximately 34 adenosine triphosphate molecules and H_2O from the carrier molecules as shown in **Figure 1**.

In this process, the electrons are transferred from bacteria to anode electrode through following mechanisms; (i) through redox-active proteins (ii) through soluble electron shuttling molecules (iii) through conductive pili and, (iv) through direct interspecies transfer mechanism [10]. However, the most dominant mechanism is assigned for electron transfer via conductive pili because they behave and show metal-like conductivity as shown in **Figure 2.**

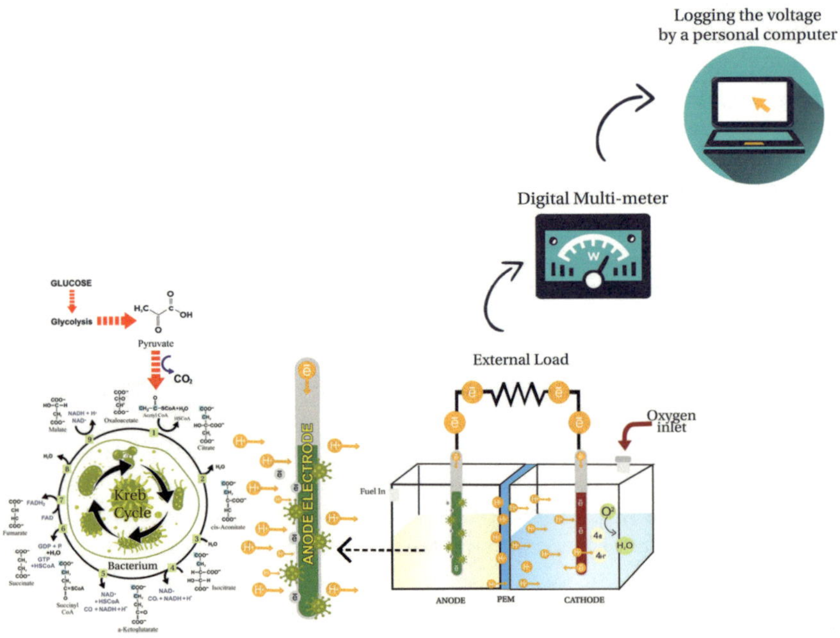

Figure 1.
Mechanism of electron transfer from bacterial cell to anode. (Adapted from Yaqoob et al. [8] reference with Elsevier permission.)

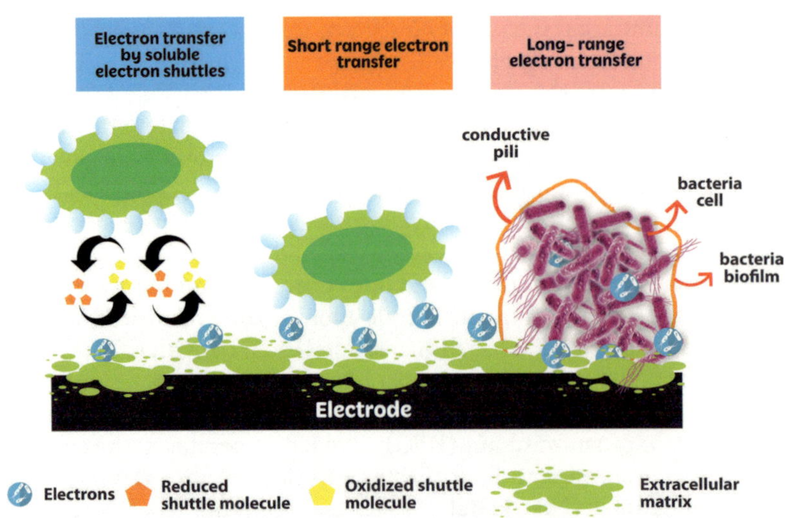

Figure 2.
Mechanism of electron transfer from microbes to anode electrode. (Adapted from Yaqoob et al. [8] reference with Elsevier permission.)

3. Required properties of anode

One part which could be significantly studied to analyze the energy generation output from MFCs operation, is the anode material. The selection of material for an ideal anode is also a promising challenge in terms of electron generation, transformation, and bacterial adhesion. However, for an ideal anode, it must have few significant properties. Many studies are present to increase the energy production by using several types of material such as conventional carbon-based materials, metal-based and conducting polymer or composite etc. [18, 19]. Although the selection criteria for the anode are different from cathode electrodes because anode electrodes are in direct contact with microbes. Therefore, some important parameters are summarized here as follows.

3.1 Electrical conductivity

The electrical conductivity of the material is considered as the most essential factor to serve as anode in MFCs operation. During operation, the electrons and protons were released by microbes which resulted as a consequence of respiration. Different mechanisms were employed to transfer the electrons from microbes to anode as shown in **Figure 2.** Three methods are essentially employed to transfer the electrons from microbes to anode (i) direct electron transfer via conductive pili (ii) electron transfer through redox-active proteins molecules (iii) use of electron shuttles to transfer electrons. However, released electrons moved from anode to cathode via an external circuit. The high conductive material helps to increase the flow of electrons and exhibit less resistance. The highly electrically conductive material reduces the resistance, which is according to Ohm's law, current is inversely proportional to resistance. The rapid flow of electrons is, because high conductive material provides a chance to release electrons to remain closer to the nucleus which generates a band where material acts as an open highway. While simultaneously, the interfacial impedance must be less between analyte and electrode to enhance the electron transfer during the process [14, 20].

3.2 Porosity and surface area

The surface area of the anode electrode carried out a direct effect on the performance of MFCs. The large surface area can offer better opportunities to microbial species for their growth and respiration effectively at the surface of anode. The anode's material porosity effects the catalytic activities of the microbes. The microbes were successfully immobilized on the surface of the anode material and produce the electron via oxidation of organic substrate. The electron harvesting was carried out in presence of ohmic losses. The internal resistance and ohmic losses of the system could be reduced by improving the anode surface area. The internal resistance is directly proportional to the ohmic losses of MFCs, so by increasing the surface area, the resistance will be reduced as stated by Chuo et al. [10]. The larger surface area of material providing a smooth space for microbes to grow and respiration occurred effectively [21]. The electrodes porosity must be sustained during operation of MFCs. However, the power generation greatly depends on the surface area of anode although high porosity in anodic material reduces the conductivity of anode material [21]. Additionally, a high surface area offers a more active site, which increases the electrode kinetics. Although many conventional carbon-based materials show better surface area, in modern trend graphene and its derivatives exhibit higher surface area than conventional carbon-based material.

3.3 Biocompatibility

The biocompatibility of anode is also very important in MFCs operation for better outcomes in terms of energy production. The produced microbes are in direct contact with the surface of anode. If the anodic material is not biocompatible with microbial growth, then the generation of electrons will be decreased. Moreover, there are several anode materials which show cytotoxicity and might reduce the microbial healthy growth on the surface of the material. The substantial potential and energy losses carried out due to absence of healthy compatibility of microbes with the anodic electrodes [22, 23].

3.4 Stability of electrode

The oxidizing and reducing atmosphere may lead to decomposition and distend of anode in MFCs. Although, the excellent roughness of surface increases durability while it might enhance the probabilities of fouling, thus may reduce the long-term stability of material. Therefore, both types of electrodes (anode & cathode) must highly show durability and stability in both the basic and acidic nature of the environment. In comprehensive, anode material always has direct contact with inoculated microbes and organic substrates which might lead to higher damage of the anode. It entirely disturbs the physical integrity of anode. Therefore, the hydrophobic-based material is preferable to minimize this mentioned effect. The electrode's stability is constrained through moieties which are engaged in material pores and reduce the space for microbial growth. The rough surface of electrodes is preferred for detaching the H_2O molecules and giving more area to microbes for their sustainability. However, optimized surface roughness and toughness of electrodes are required to enhance the performance of MFCs and minimize the adverse fouling effect [21, 24]. Furthermore, chemical, thermal, and physical stability are also important parameters for an ideal anode to work more effectively.

3.5 Availability and cost of electrode

The cost and easy availability of material for electrodes are also very important parameters and play an important role in selection of anode material. The availability with lower cost along with all above mentioned properties are considered as an ideal material for electrodes. The cost and availability of materials make the MFCs unsuitable at large scale. For practical applications of MFCs, low cost, highly stable and easily accessible materials are required. For example, nanocomposites consisting of Pt, Au etc. are highly expensive and non-sustainable materials. However, metal oxide composite with other materials such as carbon-based polymers might be a substitute to decrease the cost of material. The availability issue can be overcome by synthesizing the metal oxide and carbon-based material through green synthesis methods by using waste material. Moreover, green synthesized composites also exhibit good biocompatible, better stability and easy availability which helps to enhance the life of MFCs operation [25].

4. Material for anode electrode

The performance of MFCs and economic possibility are vitally related with progresses in anode materials. The anode electrodes have expressively impacted on formation of biofilm and transfer of electrons between electron acceptor and microbes. Several types of materials are reported earlier for the fabrication of anode for MFCs and achieved significant results, due to having the excellent properties as mentioned-above [4, 5, 9, 11]. Most familiar and used material is carbon-based, metal/metal oxide-based and conducting polymers which carry substantial outcomes, and some recent data are summarized in **Table 1.**

4.1 Carbon-based anode material

The carbon derivatives are most often used as anode electrode due to their high conductivity, excellent electrons transfer kinetics, good biocompatibility, reasonable chemical activity, mechanical and thermal stability. The carbon derivatives offer a good prospect to enhance the performance of MFCs with minimum/economic cost. Mostly used carbon derivatives are reticulated vitreous carbon, carbon paper, glassy carbon, carbon rod, carbon felt, carbon fiber, carbon cloth, activated carbon cloth, carbon brushes, graphite felt, graphite paper, 3D graphite, graphitic granular and graphene oxide powder [45]. Recently, graphite and its derivatives such as graphite rod, plate, sheet, cloth, or paper are commonly used because graphite material is more valuable than simple carbon types. The graphite material is very rigid, brittle, thin and so far, non-toxic material. The graphite material got much attention before, but later the scientific community found some drawback of this material which makes the MFCs unsuitable at higher scale i.e., low surface area and high porosity. These factors inhibit the healthier microbial growth on the surface of anode to generate the electron more rapidly [46]. Similarly, the same disadvantage is reported in case of simple carbon sheets and carbon cloth-based anodes. However, the activated carbon cloth and carbon fiber offer much better results than other conventional carbons due to reasonable surface area and better adsorption at anode [47]. Wang et al. [48] studied the carbon mesh as anode in MFCs which is cheaper than commercial carbons and exhibited reasonable current density. Zhu et al. [49] studied the graphite felt as anode and achieved the 385 W m^{-3} power density which was considered much better due to the effect of organic substrate. Therefore, there is still not enough energy efficiency to take the MFCs to industrial scale. Furthermore, carbon was employed in packing form to increase the electrode surface area for better growth of microbes [50]. The graphitic material also available in packing form and graphite packed granules showed low porosity in material but suffered from clogging which may be projected as another problem in this material [51]. Furthermore, the results were also based on charge storage efficiency and energy generation by using a single carbon-based granule. For example, the activated carbon granule stored the charges in electric double layer form which usually corresponds to enhancing the anode performance. The activated carbon-based granule produces 0.6 mA current by acting as anode. The discharge/charge mechanism shows that the activated carbon granules produce 1.3–2 times extra charge than graphitic granules. Furthermore, Zhang et al. [52] also studied graphite brushes as an anode with specific 5cm^2 diameter. The achieved power density was 1430 mW/m^2 which is better than conventional carbons anodes. Yazdi et al. [53] extensively reviewed the literature of the carbon-based material as anode electrode for MFCs. The literature showed that the modern carbon material known as carbon nanotubes (CNTs) and its modification showed cellular toxicity against the microbial community during operation. On the other hand, the carbon allotrope called graphene and its derivatives have attracted much attention recently as electrode material in MFCs. The graphene derivatives showed excellent biocompatibility, flexibility, conductivity, mechanical robustness, specific surface area, chemical inertness and stability. The commercial graphene derivatives such as graphene oxide is expensive to use as anode but according to Huggins et al. [54] it was less expensive comparatively to waste-derived graphene derivatives.

Type of Material	Electrodes		Size of Anode	Surface area of Anode (cm^2)	Power Density ($mW\ m^{-2}$)	Reference
	Anode	Cathode				
Carbon-based Anode	Plain Carbon paper	Carbon paper	2.5 × 4.5 cm	—	33	[26]
	Graphene oxide	Carbon paper	2 cm × 1 cm	—	102	[27]
	Graphene oxide wit CNT	Carbon cloth	—	—	434	[28]
	Graphene nanosheet coating on carbon paper	Carbon cloth	—	5.4	610	[29]
	Carbon cloth	Carbon cloth	2 cm × 2 cm	4	679.7	[30]
	Carbon felt	Carbon fiber felt	2.5 × 2.5 cm	2.5	784	[31]
	Graphite brush	Carbon cloth	3 cm × 2 cm	8	1280	[21]
	3D Graphene	Carbon cloth	Anode: 30 mm × 5 mm (diameter × thickness)	9.41	1516 ± 87	[32]
	Graphene	Carbon cloth	2 cm × 2 cm	4	2850	[33]
Metal and metal oxide Anode	Titanium rod	graphite felt	20 mm	20 ± 1	—	[34]
	Titanium	—	—	4	700	[35]
	Ti/TiO_2	Pt meshes	—	—	2317	[36]

Type of Material	Electrodes		Size of Anode	Surface area of Anode (cm^2)	Power Density ($mW\ m^{-2}$)	Reference
	Anode	Cathode				
Conducting Polymer & Composite-based Anode	Graphene powder/ Polytetrafluoroethylene on Carbon cloth	Carbon cloth	$4 \times 4\ cm^2$	—	0.329	[37]
	Zero-dimension nitrogen-doped carbon dots modification with carbon paper	Carbon paper	$2.5\ cm \times 2.5\ cm^2$	—	0.329	[38]
	Nickel foam/CNTs/PANI	carbon cloth	—	1 cm^2 of anode surface-area	113	[39]
	Graphene/Au composite	Carbon paper	—	6	508	[40]
	PANI networks onto GNRs-coated on carbon paper (CP/GNRs/PANI)	Carbon paper	$2\ cm \times 2\ cm$	4	856	[29]
	rGO/PPy	Carbon paper	$1\ cm \times 1.5\ cm$	—	1068	[41]
	PPy/graphene oxide	Carbon felt	$3.0\ cm \times 2.0\ cm \times 0.5\ cm$	—	1326	[42]
	rGO/ Carbon cloth-PANI	Carbon felt	$1.8\ cm \times 1.8\ cm$	—	1390	[43]
	rGO/SnO_2/Carbon cloth composite	Pt rode	$3\ cm \times 2\ cm$	6	1624	[44]
	TiO_2 and r GO composite	Carbon fiber/brush	$1\ cm \times 1\ cm$	1	3169	[21]

Table 1.
Summary of anode material with different size and surface.

The conclusion demonstrated that the biomass-derived material as anode can offer efficient performance in MFCs at reasonable cost. At the moment, to synthesize the graphene derivatives from waste materials, one of the most commonly used methods was Hummer methods [55]. However, so far very little effort seems in this direction to use waste material as electrodes for MFCs as shown in **Figure 3.**

4.2 Metal/metal oxide-based anode

The metal/metal oxide can offer high electrochemical performance in MFCs as anode, but metal corrosion limits the applications to use as anode. Silver, nickel, titanium, gold, copper, copper, aluminum etc. all metallic strips can be serving as electrodes and offer better outcomes. Metallic material carries a higher rate of conduction than other materials and thus, metal-based materials flew the electron faster which help to enhance energy generation [56–58]. The metallic electrode also showed high electrical conductivity, good mechanical stability and so far, biocompatibility for a short time. For example, Yamashita and Yokoyama, [59] studied the effect of molybdenum as in case of anode and achieved 1296 W m^{-2} power density. Metal has exclusive and remarkable properties, but still it is not broadly employed as electrode due to corrosion effect and absence of strong bacterial adhesion. The metal-based electrodes showed very less biocompatibility toward microbes, after some time of operation the metal started getting corrosion which is a serious threat to microbe's life. Nitisoravut et al. [60] studied the stainless steel as anode to enhance the energy out but after some time the achieved power density was 23 mW m^{-2} due to presence of poor microbial adhesion. Moreover, diverse dimensional biocompatible aspects of metal oxides such as vertical 3D porous structure of metal oxide (TiO_2) sheets, can offer a great active surface area, which increases the electron movement and molecular diffusion during MFCs operation. To some extent, the metal oxide offers a better electron transfer rate and good microbial adhesion. Firdous et al. [61] studied the TiO_2 based anode rod to treat the vegetable oil industries wastes and produced 5839 mV maximum voltage with 90% removal of chemical oxygen demand. The metal or metal oxide composite with other materials is a promising technique to overcome many types of issues such as cost, biocompatibility, stability, and conductivity. There are several methods to synthesize

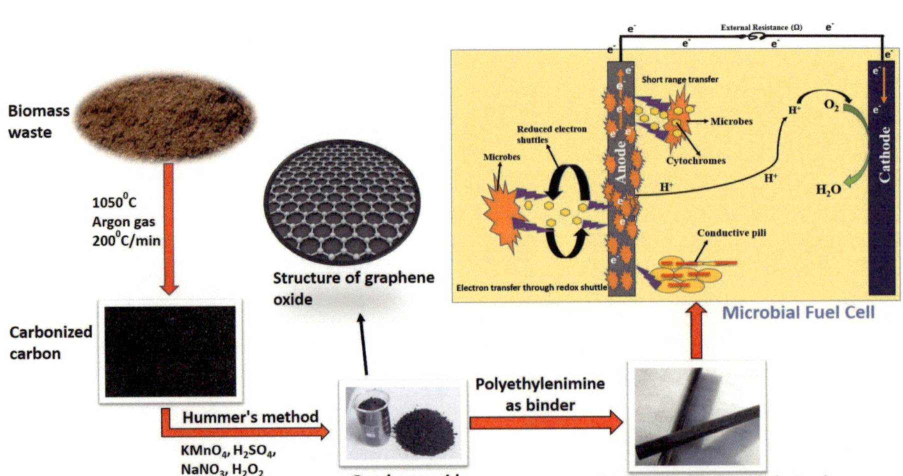

Figure 3.
Systematic demonstration of graphene-based anode fabrication for MFCs by using waste. (Reproduced from Yaqoob et al. [18] with Elsevier permission.)

the metal oxide by using the green synthesis method without any environmental hazard [21, 24]. So, a better idea would be to synthesize the metal oxide by using waste material for fabrication of anode and preferably to prepare composite with carbon-based material to enhance the performance of MFCs.

4.3 Conducting polymer material

The conductive polymer such as polycarbazole, polyaniline, poly-co-o-aminophenol, polypyrrole, polythiophene, etc. also one of the sources to fabricate the anode. Can act as anode due to having excellent electrical conductivity properties. The conductive polymers are providing significant outcomes through modification processes with other materials such as carbon-based, metal/metal oxide. For example, the polyaniline modification with carbon cloth showed better energy output as compared to unmodified material [62]. Similarly, Pandit et al. [63] studied the polyaniline coating on graphite felt and employed as anode to achieve 2.9 W m^{-3} power density, the obtained result was not good due to lack of other parameters such as organic substrate, concentration etc. The polypyrrole was considered as a potential material which exhibited 452 mW m^{-2} power density when modified on the surface of carbon paper [64]. The polypyrrole can enter a microbe cell membrane and transfer the electron by employing a metabolic pathway. Therefore, conductive polymer composites can bring a great revolution to increase the efficiency of anode. Dumitru et al. [65] studied the polyaniline/CNTs and polypyrrole/CNTs nanocomposites as anode. The CNTs/polypyrrole and CNTs/polyaniline nanocomposite showed 167.8 mW m^{-2} and 202.3 mW m^{-2} which is higher than unmodified CNTs. The observed pure CNT power density was 145.2 mW m^{-2}. The conclusion showed that the polymer modification with CNTs decreased the cellular toxicity effect of CNTs toward the microbial community. This reason led to higher energy efficiency of MFCs. The synergistic effect of CNT/conducting polymers shows high electrochemical applications. This is an emerging and promising research direction to fabricate the polymeric-based composite as anode in MFCs. Some mostly utilized metal-based, carbon-based, and conductive polymer-based materials are shown in **Figure 4.**

3D, Three dimensional; CNT, Carbon nanotubes; r GO, Reduced graphene oxide; PANI, Polyaniline; PPy, Polypyrrole; Pt, Platinum; GNRs, Graphene nanoribbons; CP, carbon paper; Ti, Tanium; TiO_2, Titanium oxide; SnO_2, Tin oxide.

Figure 4.
Mostly used anode electrodes (1) carbon paper (2) carbon mesh (3) graphite rod (4) platinum mesh (5) carbon brushes (6) carbon felt (7) carbon fiber (8) reticulated vitrified carbon (9) graphitic granular (10) different metal electrode strips. (Adapted from Yaqoob et al. [17] with MDPI permission).

5. Characterization tools

The study of MFCs can be performed by using different types of analytical, surface, biochemical, spectroscopic, and bioelectrochemical characterization methods [66]. The morphological (surface) properties analyzed characterization methods, for example, fluorescent microscopy, scanning electron microscopy (SEM), confocal laser scanning microscopy (CLSM) are very convenient in investigation of the microbial biofilm growth around the surface of anode. To understand the biofilm density, growth patterns, thickness, size and heterogeneity of the biofilm, electrochemical cauterizations are required. They are also suitable in reviewing the electrode porosity and membrane-based materials which were used in MFCs. The organic substrate application, product configuration and extracellular-based mediators in the MFCs and anolyte can be sensed and studied by a diversity of spectroscopic, biochemical, and electrochemical approaches [67]. The biochemical approaches e.g., redox mediator assesses are measurable and assistance in measurement the redox mediator's concentration, while spectroscopic systems such as HPLC, LC–MS and UV–VIS, are also vastly valuable in classifying and then computing the chemical classes. On the other hand, the electrochemical methods like differential pulse voltammetry (DPV), linear sweep voltammetry (LSV), cyclic voltammetry (CV), are qualitative techniques. They help in classifying the mechanisms which are liable for electron shuttling in current peaks form on their changeable electrochemical actions. The CV has been widely used in biological-fuel cells characterization where the mediated electron transference is the main mechanism of electron transportation. The *Shewanella sp.* secretion (flavins) and the extracellular electron transference have been acknowledged via CV and their meditations were then calculated by means of chromatographic techniques. Gradually, study based on alternative electrochemical methods, specifically electrochemical impedance spectroscopy (EIS), is today being employed. In voltammetry, through imposing possible phases or curves, the electrode is determined to a disorder from symmetry and the answer will be detected which a fleeting signal is frequently. Though, in EIS, the scheme is disconcerted with an irregular current of minor magnitude and the method follows the steady state. Additionally, the important benefit of impedance analysis is that the method is non-destructive or non-intrusive [68]. It can be achieved in the working MFCs without troubling the arrangement, while other characterization apparatuses deliberated above need the MFCs to be concerned and the products are being calm for succeeding analyses. Later, EIS is investigated in a stable state of bioelectrochemical method wherever the analyses are done without changing the current–voltage characters of the fuel cell system [69]. EIS has also been extensively studied in different zones of bioelectrochemical investigation such as fuel cells and corrosion. Moreover, EIS is a valuable tool to examine the influence of diverse internal resistances to the total impedance of the MFCs.

6. Current challenges and future perspectives

The MFCs received a lot of attention in the modern era and offered significant results. There are some modern challenges which still need attention for high electrochemical performance. The utilization of conventional electrodes has failed to improve the electric efficiency in terms of MFCs. There are various challenges related with anodes which create a hurdle in the case of MFCs applications at a commercial scale:

i. Anode materials should be economically stable for MFCs practical applications. So, making a cost-effective anode material is considered a modern challenge. For example, the high conductive graphene can offer significant results, but the cost of graphene exceed the economic feasibility of the approach. To address this problem, the effort should be on waste derived highly conductive materials (such as graphene, its composite with metal oxides/polymers) and formulate it according to the required anode design/architecture. The waste-derived resources are a virtuous source for fabricating the modified updated carbon-based anode [70].

ii. The anode development also faced another issue which is related to the binder material. It is very significant for material fabrication in any required shape. The binder selection is an important step for a researcher because this material is used to improve the properties of material, as better cohesive and firm. It is necessary to find further appropriate and economically favorable binders for development of the anode electrodes [71]. The binder played a vital role in composite anode because it was the capability of binder which binds the two materials very effectively and as a result enhanced the electron transportation rate. According to our knowledge, there is no wide-range study on the material as binders for fabrication of electrodes.

iii. The design (foam, sheet, brushes, rod, cloth,), size (length and width) of anode electrodes are very significant parts in the fabrication process. The enough surface area, conductivity, electrode spacing, of anode electrode are answerable for electron transportation from the anode in MFCs. The proper designing according to the MFCs setup drawing is very critical. The proper and feasible size and design can bring to the effective outcomes [72]. The high surface area can offer better opportunities to microbes for their respiration.

iv. Furthermore, to enhance the conductivity for the anode electrodes, the modification steps are very important in MFCs particularly in improving the power generation. Though, the appropriate mechanisms are still indistinguishable. Researchers must find a further appropriate mechanism for alterations in anodes which can be carried out more professionally. The modification refers to the composite-based anodes i.e., graphene/metal oxide, graphene/polymers, metal oxide/polymers, CNTs/metal oxide, CNTs/polymer [73]. The modern composite anode can be ideal to overcome the electron transportation issue and can bring the MFCs to the commercial status.

v. Additionally, the anode long-term stability is one of the important aspects which decreases the chances of MFCs implementation. Usually, some MFCs setup are taking a very long operation such as 4–6 month where the anode should be stable to work properly until the end of operation. The stability losses might be due to the less stable anode material or utilization of less efficiency binder materials [73–76]. Many researchers already reported that the stability of anode is a challenging step but still so far there is no proper mechanism studied to overcome this issue.

7. Conclusion

This chapter highlights the importance of anode electrode for MFCs. The anode electrodes are considered as a main fragment of MFCs approach and for survival of the biocatalyst such as microbial communities. For anode electrode development, the important growth is microbial attachment which is usually referred to as biofilm development. Substantial energies have been engrossed to raise the electrode surface area and conductivity. The high-tech electrode materials lead to the high biofilm densities. Additionally, there is still a lack of mechanism for proper development of the advanced anodes. Electrode materials are essentially very constant in the wastewater atmosphere in the MFCs process for long-term. The highly conductive materials such as graphene derivatives, metal oxides composite which can be used as anode electrode for better electrochemical performance of MFCs. Currently, the anode materials cost problem and indistinct mechanism of modification delays the modification approach in MFCs. Hence, cost effective, proper availability of materials and effective approaches for metal, or polymers nanocomposite or carbon-based type anode electrodes should be familiarized to implement MFCs applications.

Acknowledgements

The authors are gratefully acknowledged the Universiti Sains Malaysia for providing the literature review facilities.

Conflicts of interest

The authors declare that they have no conflicts of interest.

Author details

Asim Ali Yaqoob, Mohamad Nasir Mohamad Ibrahim* and Khalid Umar
School of Chemical Sciences, Universiti Sains Malaysia, Minden, Penang, Malaysia

*Address all correspondence to: mnm@usm.my

References

[1] Yaqoob AA, Parveen T, Umar K, Mohamad Ibrahim MNM. Role of nanomaterials in the treatment of wastewater: A review. Water 2020, **12**(2): 495.

[2] Logan BE, Hamelers B, Rozendal R, Schröder U, Keller J, Freguia S, *et al.* Microbial fuel cells: methodology and technology. Environmental science & technology 2006, **40**(17): 5181-5192.

[3] Nikhil G, Chaitanya DK, Srikanth S, Swamy Y, Mohan SV. Applied resistance for power generation and energy distribution in microbial fuel cells with rationale for maximum power point. Chemical Engineering Journal 2018, **335**: 267-274.

[4] Cai L, Zhang H, Feng Y, Wang Y, Yu M. Sludge decrement and electricity generation of sludge microbial fuel cell enhanced by zero valent iron. Journal of Cleaner Production 2018, **174**: 35-41.

[5] Yaqoob AA, Khatoon A, Mohd Setapar SH, Umar K, Parveen T, Mohamad Ibrahim MNM, *et al.* Outlook on the role of microbial fuel cells in remediation of environmental pollutants with electricity generation. Catalysts 2020, **10**(8): 819.

[6] Hung Y-H, Liu T-Y, Chen H-Y. Renewable coffee waste-derived porous carbons as anode materials for high-performance sustainable microbial fuel cells. ACS Sustainable Chemistry & Engineering 2019, **7**(20): 16991-16999.

[7] Yaqoob AA, Mohamad Ibrahim MNM, Umar K, Bhawani SA, Khan A, Asiri AM, Khan MR, Azam M, AlAmmari AM. Cellulose Derived Graphene/Polyaniline Nanocomposite Anode for Energy Generation and Bioremediation of Toxic Metals via Benthic Microbial Fuel Cells. Polymers 2021, **13**(1):135.

[8] Yaqoob AA, Ibrahim MNM, Yaakop AS, Umar K, Ahmad A. Modified Graphene Oxide Anode: A Bioinspired Waste Material for Bioremediation of Pb^{2+} with Energy Generation through Microbial Fuel Cells. Chemical Engineering Journal 2020, **11**:128052.

[9] ElMekawy A, Hegab HM, Losic D, Saint CP, Pant D. Applications of graphene in microbial fuel cells: The gap between promise and reality. Renewable and Sustainable Energy Reviews 2017, **72**: 1389-1403.

[10] Chuo SC, Mohamed SF, Mohd Setapar SH, Ahmad A, Jawaid M, Wani WA, Yaqoob AA, Mohamad Ibrahim MNM. Insights into the Current Trends in the Utilization of Bacteria for Microbially Induced Calcium Carbonate Precipitation. Materials. 2020, **13**(21):4993.

[11] Abbas SZ, Rafatullah M, Ismail N, Syakir MI. A review on sediment microbial fuel cells as a new source of sustainable energy and heavy metal remediation: mechanisms and future prospective. International Journal of Energy Research 2017, **41**(9): 1242-1264.

[12] Abbas SZ, Rafatullah M, Ismail N, Nastro RA. Enhanced bioremediation of toxic metals and harvesting electricity through sediment microbial fuel cell. International Journal of Energy Research 2017, **41**(14): 2345-2355.

[13] Wang H, Park J-D, Ren ZJ. Practical energy harvesting for microbial fuel cells: a review. Environmental science & technology 2015, **49**(6): 3267-3277.

[14] Palanisamy G, Jung H-Y, Sadhasivam T, Kurkuri MD, Kim SC, Roh S-H. A comprehensive review on microbial fuel cell technologies: Processes, utilization, and advanced developments in electrodes and

membranes. Journal of cleaner production 2019, **221:** 598-621.

[15] Fadzli FS, Rashid M, Yaqoob AA, Ibrahim MNM. Electricity Generation and Heavy Metal Remediation by Utilizing Yam (Dioscorea alata) Waste in Benthic Microbial Fuel Cells (BMFCs). Biochemical Engineering Journal 2021, **172**:108067.

[16] Chuo SC, Mohamed SF, Mohd Setapar SH, Ahmad A, Jawaid M, Wani WA, *et al.* Insights into the Current Trends in the Utilization of Bacteria for Microbially Induced Calcium Carbonate Precipitation. Materials 2020, **13**(21): 4993.

[17] Yaqoob AA, Ibrahim MNM, Rafatullah M, Chua YS, Ahmad A, Umar K. Recent advances in anodes for microbial fuel cells: An Overview. Materials 2020, **13**(9): 2078.

[18] Yaqoob AA, Ibrahim MNM, Rodríguez-Couto S. Development And modification of materials to build cost-effective Anodes for microbial fuel cells (MFCs): An overview. Biochemical Engineering Journal 2020: 107779.

[19] Yaqoob AA, Ibrahim MNM. A Review Article of Nanoparticles; Synthetic Approaches and Wastewater Treatment Methods. International Research Journal of Engineering and Technology 2019, **6:** 1-7.

[20] Hindatu Y, Annuar M, Gumel A. Mini-review: Anode modification for improved performance of microbial fuel cell. Renewable and Sustainable Energy Reviews 2017, **73:** 236-248.

[21] Santoro C, Arbizzani C, Erable B, Ieropoulos I. Microbial fuel cells: from fundamentals to applications. A review. Journal of power sources 2017, **356:** 225-244.

[22] Tan RKL, Reeves SP, Hashemi N, Thomas DG, Kavak E, Montazami R, *et al.* Graphene as a flexible electrode: review of fabrication approaches. Journal of Materials Chemistry A 2017, **5**(34): 17777-17803.

[23] Yaqoob AA, Ahmad H, Parveen T, Ahmad A, Oves M, Ismail IM, *et al.* Recent Advances in Metal Decorated Nanomaterials and Their Various Biological Applications: A Review. Frontiers in Chemistry 2020, **8:** 341.

[24] Erable B, Byrne N, Etcheverry L, Achouak W, Bergel A. Single medium microbial fuel cell: stainless steel and graphite electrode materials select bacterial communities resulting in opposite electrocatalytic activities. International Journal of Hydrogen Energy 2017, **42**(41): 26059-26067.

[25] Rahimnejad M, Adhami A, Darvari S, Zirepour A, Oh S-E. Microbial fuel cell as new technology for bioelectricity generation: A review. Alexandria Engineering Journal 2015, **54**(3): 745-756.

[26] Logan BE, Wallack MJ, Kim K-Y, He W, Feng Y, Saikaly PE. Assessment of microbial fuel cell configurations and power densities. Environmental Science & Technology Letters 2015, **2**(8): 206-214.

[27] Chen J, Deng F, Hu Y, Sun J, Yang Y. Antibacterial activity of graphene-modified anode on Shewanella oneidensis MR-1 biofilm in microbial fuel cell. Journal of Power Sources 2015, **290:** 80-86.

[28] Kumar GG, Hashmi S, Karthikeyan C, GhavamiNejad A, Vatankhah-Varnoosfaderani M, Stadler FJ. Graphene oxide/carbon nanotube composite hydrogels—versatile materials for microbial fuel cell applications. Macromolecular rapid communications 2014, **35**(21): 1861-1865.

[29] Zhao C, Wang Y, Shi F, Zhang J, Zhu J-J. High biocurrent generation in

Shewanella-inoculated microbial fuel cells using ionic liquid functionalized graphene nanosheets as an anode. Chemical communications 2013, **49**(59): 6668-6670.

[30] Qiao Y, Wen G-Y, Wu X-S, Zou L. L-Cysteine tailored porous graphene aerogel for enhanced power generation in microbial fuel cells. RSC Advances 2015, **5**(72): 58921-55892.

[31] Yang W, Kim K-Y, Saikaly PE, Logan BE. The impact of new cathode materials relative to baseline performance of microbial fuel cells all with the same architecture and solution chemistry. Energy & Environmental Science 2017, **10**(5): 1025-1033.

[32] Huang L, Li X, Ren Y, Wang X. A monolithic three-dimensional macroporous graphene anode with low cost for high performance microbial fuel cells. RSC advances 2016, **6**(25): 21001-21010.

[33] Najafabadi AT, Ng N, Gyenge E. Electrochemically exfoliated graphene anodes with enhanced biocurrent production in single-chamber air-breathing microbial fuel cells. Biosensors and Bioelectronics 2016, **81**: 103-110.

[34] Michaelidou U, Ter Heijne A, Euverink GJW, Hamelers HV, Stams AJ, Geelhoed JS. Microbial communities and electrochemical performance of titanium-based anodic electrodes in a microbial fuel cell. Applied and Environmental Microbiology 2011, **77**(3): 1069-1075.

[35] Dominguez-Benetton X, Srikanth S, Satyawali Y, Vanbroekhoven K, Pant D. Enzymatic electrosynthesis: an overview on the progress in enzyme-electrodes for the production of electricity, fuels and chemicals. J Microb Biochem Technol 2013, **6**(2).

[36] Benetton XD, Navarro-Ávila S, Carrera-Figueiras C. Electrochemical evaluation of Ti/TiO2-polyaniline anodes for microbial fuel cells using hypersaline microbial consortia for synthetic-wastewater treatment. J New Mater Electrochem Sys 2010, **13**: 1-6.

[37] Pareek A, Sravan JS, Mohan SV. Fabrication of three-dimensional graphene anode for augmenting performance in microbial fuel cells. Carbon Resources Conversion 2019, **2**(2): 134-140.

[38] Guan Y-F, Zhang F, Huang B-C, Yu H-Q. Enhancing electricity generation of microbial fuel cell for wastewater treatment using nitrogen-doped carbon dots-supported carbon paper anode. Journal of Cleaner Production 2019, **229**: 412-419.

[39] Nourbakhsh F, Mohsennia M, Pazouki M. Nickel oxide/carbon nanotube/polyaniline nanocomposite as bifunctional anode catalyst for high-performance Shewanella-based dual-chamber microbial fuel cell. Bioprocess and biosystems engineering 2017, **40**(11): 1669-1677.

[40] Zhao C-e, Gai P, Song R, Zhang J, Zhu J-J. Graphene/Au composites as an anode modifier for improving electricity generation in Shewanella-inoculated microbial fuel cells. Analytical Methods 2015, **7**(11): 4640-4644.

[41] Gnana Kumar G, Kirubaharan CJ, Udhayakumar S, Ramachandran K, Karthikeyan C, Renganathan R, *et al.* Synthesis, structural, and morphological characterizations of reduced graphene oxide-supported polypyrrole anode catalysts for improved microbial fuel cell performances. ACS Sustainable Chemistry & Engineering 2014, **2**(10): 2283-2290.

[42] Lv Z, Chen Y, Wei H, Li F, Hu Y, Wei C, *et al.* One-step electrosynthesis of polypyrrole/graphene oxide composites for microbial fuel cell application. Electrochimica Acta 2013, **111**: 366-373.

[43] Hou J, Liu Z, Zhang P. A new method for fabrication of graphene/polyaniline nanocomplex modified microbial fuel cell anodes. Journal of Power Sources 2013, **224:** 139-144.

[44] Mehdinia A, Ziaei E, Jabbari A. Facile microwave-assisted synthesized reduced graphene oxide/tin oxide nanocomposite and using as anode material of microbial fuel cell to improve power generation. International journal of hydrogen energy 2014, **39**(20): 10724-10730.

[45] Yaqoob AA, Ibrahim MNM, Ahmad A, Reddy AVB. Toxicology and Environmental Application of Carbon Nanocomposite. *Environmental Remediation Through Carbon Based Nano Composites*. Springer, 2020, pp 1-18.

[46] Katuri KP, Kalathil S, Ragab Aa, Bian B, Alqahtani MF, Pant D, *et al.* Dual-function electrocatalytic and macroporous hollow-fiber cathode for converting waste streams to valuable resources using microbial electrochemical systems. *Advanced Materials* 2018, **30**(26): 1707072.

[47] Sokol NW, Bradford MA. Microbial formation of stable soil carbon is more efficient from belowground than aboveground input. Nature Geoscience 2019, **12**(1): 46-53.

[48] Wang K, Liu Y, Chen S. Improved microbial electrocatalysis with neutral red immobilized electrode. Journal of Power Sources 2011, **196**(1): 164-168.

[49] Zhu D, Wang D-B, Song T-s, Guo T, Ouyang P, Wei P, *et al.* Effect of carbon nanotube modified cathode by electrophoretic deposition method on the performance of sediment microbial fuel cells. Biotechnology letters 2015, **37**(1): 101-107.

[50] Guo X, Zhan Y, Chen C, Cai B, Wang Y, Guo S. Influence of packing material characteristics on the performance of microbial fuel cells using petroleum refinery wastewater as fuel. Renewable Energy 2016, **87:** 437-444.

[51] Borsje C, Liu D, Sleutels TH, Buisman CJ, ter Heijne A. Performance of single carbon granules as perspective for larger scale capacitive bioanodes. Journal of Power Sources 2016, **325:** 690-696.

[52] Zhang C, Liang P, Yang X, Jiang Y, Bian Y, Chen C, *et al.* Binder-free graphene and manganese oxide coated carbon felt anode for high-performance microbial fuel cell. Biosensors and Bioelectronics 2016, **81:** 32-38.

[53] Yazdi AA, D'Angelo L, Omer N, Windiasti G, Lu X, Xu J. Carbon nanotube modification of microbial fuel cell electrodes. Biosensors and Bioelectronics 2016, **85:** 536-552.

[54] Chen S, Liu Q, He G, Zhou Y, Hanif M, Peng X, Wang S, Hou H. Reticulated carbon foam derived from a sponge-like natural product as a high-performance anode in microbial fuel cells, Journal of Materials Chemistry 2012, **22:**18609-18613.

[55] Deng Y, Fang C, Chen G. The developments of SnO2/graphene nanocomposites as anode materials for high performance lithium ion batteries: a review. Journal of Power Sources 2016, **304:** 81-101.

[56] Zerrouki A, Salar-García M, Ortiz-Martínez V, Guendouz S, Ilikti H, de Los Ríos A, *et al.* Synthesis of low cost organometallic-type catalysts for their application in microbial fuel cell technology. Environmental technology 2019, **40**(18): 2425-2435.

[57] Yaqoob AA, Umar K, Ibrahim MNM. Silver nanoparticles: various methods of synthesis, size affecting factors and their potential applications–a review. Applied Nanoscience 2020: 1-10.

[58] Yaqoob AA, Khan R, Saddique A. Review article on applications and classification of gold nanoparticles. International Journal of Research 2019, **6:** 762-768.

[59] Druzhinina IS, Kopchinskiy AG, Kubicek EM, Kubicek CP. A complete annotation of the chromosomes of the cellulase producer Trichoderma reesei provides insights in gene clusters, their expression and reveals genes required for fitness. Biotechnology for biofuels 2016, **9**(1)**:** 1-16.

[60] Nitisoravut R, Thanh CN, Regmi R. Microbial fuel cells: Advances in electrode modifications for improvement of system performance. International Journal of Green Energy 2017, **14**(8)**:** 712-723.

[61] Firdous S, Jin W, Shahid N, Bhatti Z, Iqbal A, Abbasi U, *et al.* The performance of microbial fuel cells treating vegetable oil industrial wastewater. Environmental Technology & Innovation 2018, **10:** 143-151.

[62] Shahadat M, Bushra R, Khan MR, Rafatullah M, Teng TT. A comparative study for the characterization of polyaniline based nanocomposites and membrane properties. RSC Advances 2014, **4**(40)**:** 20686-20692.

[63] Pandit S, Chandrasekhar K, Jadha D, Ghangrekar MM, Das D. Contaminant removal and energy recovery in microbial fuel cells. Microbial Biodegradation of Xenobiotic Compounds 2019**:** 76.

[64] Phoon BL, Lai CW, Juan JC, Show PL, Chen WH. A review of synthesis and morphology of SrTiO 3 for energy and other applications. International Journal of Energy Research 2019, **43**(10)**:** 5151-5174.

[65] Dumitru A, Vulpe S, Radu A, Antohe S. Influence of nitrogen environment on the performance of conducting polymers/CNTs nanocomposites modified anodes for microbial fuel cells (MFCs). Rom J Phys 2018, **63:** 605-625.

[66] Kannan MV, kumar G.G. Current status, key challenges and its solutions in the design and development of graphene based ORR catalysts for the microbial fuel cell applications. Biosensors and Bioelectronics. 2016, **77**:1208-1220.

[67] Sekar N, Ramasamy RP. Electrochemical impedance spectroscopy for microbial fuel cell characterization. J Microb Biochem Technol S 2013, **6**(2).

[68] Deb D, Patel R, Balas VE. A Review of Control-Oriented Bioelectrochemical Mathematical Models of Microbial Fuel Cells. Processes. 2020, **8(5)**: 583.

[69] Kashyap D, Dwivedi PK, Pandey JK, Kim YH, Kim GM, Sharma A, Goel S. Application of electrochemical impedance spectroscopy in bio-fuel cell characterization: A review. International journal of hydrogen energy 2014, **39(35)**:20159-20170.

[70] Yaqoob AA, Serrà A, Ibrahim MNM, Yaakop AS. Self-assembled oil palm biomass-derived modified graphene oxide anode: An efficient medium for energy transportation and bioremediating Cd (II) via microbial fuel cells. Arabian Journal of Chemistry 2021,**103121**, 1-16.

[71] Chang H-C, Gustave W, Yuan Z-F, Xiao Y, Chen Z. One-step fabrication of binder-free air cathode for microbial fuel cells by using balsa wood biochar. Environmental Technology & Innovation 2020, **18:** 100615.

[72] Sekeri SH, Ibrahim MNM, Umar K, Yaqoob AA, Azmi MN, Hussin MH, *et al.* Preparation and characterization of nanosized lignin from oil palm (Elaeis guineensis) biomass as a novel

emulsifying agent. International journal of biological macromolecules 2020, **164:** 3114-3124.

[73] Umar K, Yaqoob AA, Ibrahim MNM, Parveen T, Safian MT. Environmental applications of smart polymer composites. Smart Polym. Nanocompos. Biomed. Environ. Appl. 2020, **15**: 295-320.

[74] Cecconet D, Sabba F, Devecseri M, Callegari A, Capodaglio AG. In situ groundwater remediation with bioelectrochemical systems: A critical review and future perspectives. Environment International 2020, **137:** 105550.

[75] Asim AY, Ibrahim MNM, Khalid U, Tabassum P, Akil A, Lokat D, Siti H.A. A glimpse into the microbial fuel cells for the wastewater treatment with energy generation. Desalination and Water Treatment 2021, **214:** 379-389.

[76] Yaqoob AA, Ibrahim MN, Guerrero-Barajas C. Modern trend of anodes in microbial fuel cells (MFCs): An overview. Environmental Technology & Innovation. 2021, **23:** 101579.